Understanding IMAGES

Understanding IMAGES

FINDING MEANING IN DIGITAL IMAGERY

Francis T. Marchese

Francis T. Marchese
Computer Graphics Laboratory
Pace University
One Pace Plaza
New York, N.Y. 10038, USA

Publisher: Allan M. Wylde
Publishing Associate: Kate McNally Young
Product Manager: Carol Wilson
Production/Manufacturing Manager: Jan V. Benes
Cover Designer: Iva Frank

Cataloging-in-Publication Data is available from the Library of Congress

Photocomposed pages prepared from the author's TEX files.

9 8 7 6 5 4 3 2 1

ISBN 0-387-94148-7

TELOS, The Electronic Library of Science, is an imprint of Springer-Verlag New York with publishing facilities in Santa Clara, California. Its publishing program encompasses the natural and physical sciences, computer science, economics, mathematics, and engineering. All TELOS publications have a computational orientation to them, as TELOS' primary publishing strategy is to wed the traditional print medium with the emerging new electronic media in order to provide the reader with a truly interactive multimedia information environment. To achieve this, every TELOS publication delivered on paper has an associated electronic component. This can take the form of book/diskette combinations, book/CD-ROM packages, books delivered via networks, electronic journals, newsletters, plus a multitude of other exciting possibilities. Since TELOS is not committed to any one technology, any delivery medium can be considered.

The range of TELOS publications extends from research level reference works through textbook materials for the higher education audience, practical handbooks for working professionals, as well as more broadly accessible science, computer science, and high technology trade publications. Many TELOS publications are interdiscilinary in nature, and most are targeted for the individual buyer, which dictates that TELOS publications be priced accordingly.

Of the numerous definitions of the Greek word "telos," the one most representative of our publishing philosophy is "to turn," or "turning point." We perceive the establishment of the TELOS publishing program to be a significant step towards attaining a new plateau of high quality information packaging and dissemination in the interactive learning environment of the future. TELOS welcomes you to join us in the exploration and development of this frontier as a reader and user, an author, editor, consultant, strategic partner, or in whatever other capacity might be appropriate.

TELOS, The Electronic Library of Science
Springer-Verlag Publishers
3600 Pruneridge Avenue, Suite 200
Santa Clara, CA 95051

TELOS Diskettes

Unless otherwise designated, computer diskettes packaged with TELOS publications are 3.5″ high-density DOS-formatted diskettes. They may be read by any IBM-compatible computer running DOS or Windows. They may also be read by computers running NEXTSTEP, by most UNIX machines, and by Macintosh computers using a file exchange utility.

In those cases where the diskettes require the availability of specific software programs in order to run them, or to take full advantage of their capabilities, then the specific requirements regarding these software packages will be indicated.

TELOS CD-ROM Discs

It is also clearly indicated to buyers of TELOS publications containing CD-ROM discs, or in cases where the publication is a standalone CD-ROM product, the exact platform, or platforms, on which the disc is designed to run. For example, Macintosh only; MPC only; Macintosh and Windows (cross-platform), etc.

TELOSpub.com (Online)

New product information, product updates, TELOS news and FTPing instructions can be accessed by sending a one-line message: `send info` to: `info@TELOSpub.com`. The TELOS anonymous FTP site contains catalog product descriptions, testimonials and reviews regarding TELOS publications, data-files contained on the diskettes accompanying the various TELOS titles, order forms and price lists.

Contents

Understanding IMAGES

1

Editor's Introduction

Francis T. Marchese[1]

From Paleolithic cave paintings at Lascaux depicting the hunt, to computer constructed fly-bys over canyons of Venus, imagery remains essential to human communication. Pictures, demonstrations, music, and dance have played significant roles in moving information from one mind to another. The use of computers as expressive and visualization tools has enhanced the ability of artists, designers, scientists, engineers, and educators to further explore and communicate complex concepts through the use of computer graphic based hypermedia, multimedia, and virtual reality. These electronic media radically amplify each step of the image construction, transmission, decipherment, and understanding process. And since the seeing process is strongly influenced by human anatomy and physiology, education and culture, electronic communicators have begun to ask: *What are the fundamental standards and modes of communication for creating comprehensible and meaningful electronic imagery* [1]-[3]*?*

In an attempt to define and discuss issues essential to image understanding within the computer graphics context, a conference entitled Understanding Images was held at Pace University's Manhattan campus on May 21-22, 1993. Sponsored by the New York City local group of ACM SIGGRAPH and Pace University's School of Computer Science and Information Systems, over 200 computer scientists; traditional, contemporary, and electronic artists and designers; architects, photographers, philosophers, psychologists, cognitive scientists, and musicians from the United States and Europe gathered for two days of spirited presentation, discussion, and debate.

Fourteen papers given at the conference are found here in the order of presentation. The breadth of topics fall into general categories of text, sound, design, image analysis, psychology, and perception. But the conference's true richness and complexity results from the

[1]Computer Science Department, Pace University, NY, NY 10038-1502.

cross-disciplinary perspectives of its participants. For example, Marc de Mey, a psychologist, used threedimensional computer graphic rendering to reconstruct Masaccio's *Trinity*, in order to understand this early Renaissance painter's use of perspective; Thomas Hubbard, a photo journalist, engaged the entire audience in interactive image analysis; or Robert Williams, a computer scientist and musician, who performed Mussorgsky's *Pictures at an Exhibition* on a computer assisted classical guitar, demonstrating a guitarist's ability to translate simple gestures for enhanced sonic expression.

Each paper in this proceedings is a point on the perceptual horizon. Together they form a Gestalt that organizes the visual landscape, highlighting reconverging paths of art, science, and technology toward a future Renaissance. Just as the fifteenth century Renaissance painter reinvented the technology of perspective to organize and represent images, future Renaissance communicators will reinvent computer technology to help us better understand visual reality.

This conference would have not been successful without support from a few dedicated individuals. Specifically, I would like to thank Jean Coppola (conference coordinator), the NYC ACM SIGGRAPH board of directors, Dr. Carol Wolf (Chair, Computer Science Department), Kenneth Norz (Assistant Dean), Dr. Susan Merritt (Dean, School of Computer Science and Information Systems), and the director and staff of Pace University's Downtown Theater.

1.1 References

[1] Barlow, H., Blackmore,C. , and Weston-Smith, M. (Eds.) (1990). *Images and Understanding.* Cambridge:Cambridge University Press.

[2] Ellis, R.E., Kaiser, M., and Grunwald, A.J (Eds.) (1991). *Pictoral Communication in Virtual and Real Environments.* London: Taylor and Francis Ltd.

[3] Tufte, E.R. (1983). *Visual Display of Quantitative Information.* Chesire, CT: Graphics Press.

2

Lipstick on the Bulldog

Alyce Kaprow[1,2]

2.1 Introduction

Designers define, articulate, synthesize, collaborate, create, implement, build consensus, organize, compromise, communicate, visualize and make something real by putting a "casing" around the "product."

Steve Jobs once said he envisioned the computer as prolific as the toaster. Nicholas Negroponte predicted that there would be a digital clock in every appliance in the home and office. Both were right.

As the industries that produce computers and consumer electronics mature, and as their once-rarefied products become commodities, we face crossroads. No longer are the primary users of computers and electronics necessarily people with professional degrees in computer science and engineering; they are people using everyday tools for everyday tasks at home, in public arenas, and in business.

As products demand more and more from the users because they deliver more and more complicated functionality, the role of design increases in importance. As the functions of similar products mimic one another, the manner and ease in which they operate will define their excellence and distinguish them from their competition.

Our products must meet the needs of a multi-cultural, multi- lingual, multi-skilled, and multi-goaled population. We cannot assume that the user will be fluent in computer-speak, patient with the inconveniences of version-one solutions, and delighted over each and every technical breakthrough. Some will be technophobic, some impatient, some physically challenged, some illiterate. Athletes, doctors, teachers, laborers, computer programmers, CEO's, farmers, all

---

[1]The New Studio, 26 Hope Street, Newton MA 02166

[2]Original version of this document appeared as introduction to the Designing Technology show catalog in the Visual Proceedings of the SIGGRAPH '93 Conference

use a diversity of products. They are for all of us; and the coherence of their look and feel will make or break their success. All products were "designed," though some not as well as others.

Designers play a critical role in product development, and a multidisciplinary collaborative team better insures product excellence. The message from those who understand the necessity of coherent and collaborative design, and the value of design-thinking, is that designers not only improve the visual appearance, but bring a coherence and logic to the planning that is unique and extremely important.

Interaction designers and design strategists are advocates for the user. They define and map functionality, navigation, and usability. Designers combine and synthesize, visualize and communicate, and lay out 2-D, 3-D, and virtual space. They bring to product planning coherence and an overview of these issues.

Designers contribute to the development effort differently than those that come to it from the engineering or business/marketing sides. Most design discussions are more about functionality and usability than style and appearance. A product's "look" evolves so that its usefulness and usability are improved. What may appear to some as an exercise in decoration is really about how the product should function.

The message this show is sending is that collaborative teamwork that includes the designer from day-one will significantly improve the product, and will add value to the end result – not just stylistic decoration. The development process will make the difference; excellent products are the proof of its success.

2.2 A Few Definitions

Let's begin with three definitions as they pertain to the theme of *Designing Technology*:

- **Product** is anything that results from a development effort. That can be a three- dimensional object, a space or environment, printed material of any size and shape, or the virtual that exists only on a screen. It can be a complex appliance, a bicycle, an operating system, a user's manual, an on-screen help system, or a multimedia presentation.

- **Design** refers to organizing and styling information. It includes graphic design, surface and volume design, spatial design, and screen design that define the product's look and feel. Design also considers the interaction, content, communication and temporal relationships. It is the coordination between all of these to present a coherent package.

- **Interface** is the "dialog" between product and user. Interface is about interaction, and interaction happens between all types of users and all types of products.

2.3 Why Collaboration is Hard to Do

It is commonly taught that work is best done in an isolated, singular manner. We sat at our school desks and were instructed never to discuss answers and ideas with others in the class. Those that came to the solution first and alone were rewarded. Only in some rare instances do we now find project-oriented classrooms and curricula.

This isolated approach is continued in the work place, especially by those who excelled in school using these methods, often times not realizing that they send a message contrary to collaboration and teamwork. The hierarchical relationships that result from this methodology perpetuate the myth that collaboration on an equal ground is the antithesis of individual excellence. Well, this just doesn't work very well in our multigoaled, multicultural, multiskilled, multilingual world. Those who have experienced the collaborative approach invariably understand that synergy and strength in numbers and insight will create a solution far greater than the individual parts. Those that understand this not only have insight into the expertise of other disciplines (thus a heightened awareness of the product's requirements), but also gain an inherent trust and respect for all team members' contributions.

Collaboration is hard because it often counters the way we have learned to achieve. It requires a different standard for evaluating the progress of development. Participatory design, where teams evolve solutions based on user-centered analysis, works. The presentations in the show prove this.

There are at least two different ways to analyze a problem: with a step-by-step analysis of each task that needs to be done (the trees);

and by seeing the overall project as a coordinated effort (the forest). Both need to be part of the analysis, but one approach may not be clearly seen by the other. Additionally, people who excel in creating and caring for the individual trees may not choose to worry about the design of the whole forest; and vice versa. Collaboration is good and brings together these methods and skills. Project managers in charge of development need to understand both points of view, but often excel in only one.

I believe that it is far easier to articulate the measurable tasks between point A and point B than it is to articulate the subjective overview of how work is progressing holistically. There are parallels between this analysis and that of design and engineering. It is often difficult to justify something that is not specifically measurable, like design and the psychology of human factors; but such contributions are essential.

The process of design is often thought of as mysterious and vague. It is not always specifically quantifiable; it does not always have measurable points of demarcation; the process is more subjective and holistic. At times it is a lack of understanding about what design is that causes the confusion.

2.4 Process, Tools, Product

For the purpose of description, it is convenient to look at the topic of development in three areas: process, tools, and product. The process is the continuum of steps and stages, combined with people and ideas that shape the product's outcome from concept to finalization. The product is the result of the effort, as described earlier. The tools are those used by the developers to describe, plan, and manufacture the product. In the context of collaborative development, the process is the most important of the three.

The goal of producing excellent products should be a given. While this goal is clearly not always achieved, for the sake of convenience, let's work backwards and begin with excellent products. How to guarantee this is precisely the issue at hand.

Tools describe the method of operation and make the process possible. A designer can sketch a product drawing with a pencil, create a prototype in clay or foam-board, or describe it with a CAD workstation. What tools are used to describe, plan, and manufacture the product will be chosen because they are available, dexterous, comfortable, able to communicate the essential information, and accessible. A stick in sand is as valid a tool as a million-dollar workstation, so long as it is appropriate and communicates information to those who need it.

The choice of tools to a certain degree dictate the manner in which people work and communicate. They may even leave their mark stylistically. Tools can reduce the time we spend on tasks and shift our work styles.

The process is the key issue. Our working relationships and expectations about each other's responsibilities and contributions set the style of product development. The designer should be a key team member in this scenario because the main experiences and insights that a designer brings to the project are the abilities to analyze and visualize and to synthesize information in a structured, coherent way.

The design process is more than the ability to sketch and choose colors. Designers should not work in isolation any more than engineers should. Nor should their skills be tacked onto the end of a project to make it look good. What should be avoided is the development cycle where product specs are "thrown over the wall" from marketing to engineering to design to human factors, without the dialog between all the collaborators along the way. How can we attempt to develop for our mult faceted world without the expertise from every corner offering insight and analysis? How can we compete with those who do?

2.5 Strategists or Service Bureaus

Designers are not all alike, just as programmers are not equally agile in writing all types of software. Some are graphic designers, some design form and volume, and some design space. Many designers are concerned with individual appearance and signature, creating fabulous printed material, furniture, clothing, and other artifacts that carry a very personal message and style. Other designers focus their attention on individual pieces for clients, such as a brochure or illustration, an annual report or poster. They typically do not get involved with long-term planning for the product line or company identity.

This show is about interaction design, and designers who actively plan and define products. It is about defining product look and feel, functionality and usability, company strategy and identity.

The show challenges the perception of those who hire the skills and talents of the people in the design community. Many people consider design to be a service trade, producing items such as business graphics or a trade show booth. At best, in these circumstances, designers are thought to be a brilliant visualizers of ideas; at worst they are perceived as service bureaus. But, insight into the overall strategy and direction a company and/or product can take is also a valued design contribution. Presentations in this show illustrate a different concentration of design innovation and farther-reaching roles that are possible.

The intersection of design and technology is often represented by the tools the design community now has available: page layout, illustration, CAD, three-dimensional modeling, rendering and the like. These examples – excellent as they are – are drawn from situations involving tool users, not tool makers. Though the work that is being done might be brilliant examples of design, it does not necessarily exemplify the role of design and visualization in the process of product development.

This is a very fine-line distinction, and one that often confuses people. Indeed, the designer who produces a brilliant poster, strong in style and accent, may be the same designer who is involved with the interface of a new operating system, and may very well use the same computer tools. The focus is different for different clients. There is, happily, a transfer of skills from one activity to the other.

2.6 Education

Environmental designer Kevin Lynch maintains that 80% of the information we absorb is visual. If that is true, why do we spend so little time during our formal education learning to understand visual language and how to communicate visually? Indeed, it is worse than ignored, it is often put into the category of expendable when budgets are tight. If look-and-feel are so important to the success of products and communications, why do we continually place the skills associated with design in a category of second-class citizenship?

Visual communication needs to be taught, just as verbal communication and mathematical skills are; all are essential. This must begin in primary school with a concerted effort and respect for the value such skills bring to problem solving and communication. Scientific visualization is essential, as is the visualization of business data, the definition of volume and form, and the layout of an environment. Not everyone will become a designer; but, the respect for a designer's contribution will be a direct result of the respect visual skills are given in school.

Companies must educate their workers to understand and respect the input of disparate and collaborative groups. Though it appears to threaten the control of hierarchical structures, such creative communication is essential for creative product development. Collaborative teamwork does not mean a free-for-all without leadership. The design process is no more chaotic than any creative brainstorming; accountability and responsibility are still important. But, often such a team seems to threaten many who don't understand it. We must begin training in visual communication and collaborative problem solving in the earliest schools, and continue to reinforce these ideas in the work place.

Human factors experts are often viewed as substitutes for designers. Engineers who code the interface typically place it on the screen as they see convenient. When it doesn't work, they call in the designer to "fix" it. SOME DESIGNERS VIEW THESE ELEVENTH-HOUR ATTEMPTS AS "PUTTING LIPSTICK ON A BULLDOG". It is no substitute for good design from the start.

Programmers and product planners often ask how to work with designers. Conversely, designers ask how to work with programmers and hardware engineers. Designers visualize and have strong opinions about intangible issues; engineers may make C-code jokes and describe colors by RGB numerical equivalents. Neither starts out trusting the other. It is essential to learn how to converse, how to share common ground, and to respect the other for the quality of experience and insight and the difference in point of view. We must learn enough about each other's domains to communicate essential information on a continual basis.

Respect for differences in approach and the ability to take risks are critical to solving the needs of our world and in manufacturing worthy and reliable goods. The strength to challenge the norm and to take a risk is what must continue to drive this industry.

2.7 Common Ground

All of us are all part of the same effort, whether we come from marketing, engineering, education, design, cognitive psychology, anthropology, training, or elsewhere. The work in the show exemplifies the methodology and philosophies of collaboration and design visualization. It includes all facets of product development and invites everyone to engage in this collaborative method.

In one way or another, most products enable and empower us to do things – an expanded definition for tools. They must work well; they must be comfortable and familiar to be used safely and effectively. The telephone is a tool. It has changed the way we work and communicate, and it changed our expectations about almost everything we do. FAX machines have gone one step further. Desktop personal computers with powerful graphics, layout, and imaging software plus peripherals have empowered each of us in ways unthought of just five years ago.

Design is not a luxury for Fortune-500 companies; it is a necessity if companies are to compete in world markets. Communication on a global scale necessitates coherence and accessibility of information. What will distinguish one product from another is not the abundance of functions but the manner in which they are presented to the users.

Some of the best examples of well designed products are those described as "fun to use", "comfortable", "familiar", and "easy." They invite people to use them, they teach people to master them, and they challenge people to go further. Most are very good looking as well. We enjoy using them; this is not a negative feature, but value-added....and no accident. Well designed products help sell themselves.

Responsible design is everyone's business. This not only includes the look and feel, but its longevity, its credibility as a useful product, and its responsiveness to the environment, both visually and ecologically.

Designers need to embrace this cause as well. It is the combined responsibility of the design and engineering communities to champion collaborative and participatory development. If we need the powers of "left-brain thinkers" to define our logical world, we need the powers of the "right- brain thinkers" to handle the visual and temporal one. More importantly we need communication between them. This will happen more effectively once we allow our own brains to converse from both sides.

Mutual respect and delight for each contributor on the team is an essential beginning. Not trying to second-guess the expertise and insight of each other's disciplines will allow open discussion and collaboration. Designers have a responsibility to champion their own talents and expertise; we all have the need for dialog.

The computer is perhaps the most powerful tool we have at out fingertips; but, until it became visual, it was thought to be exclusive and unforgiving. We now have opportunities to empower people. Taking on the challenge of enriching our lives will take the concerted effort of all skills and talents, from both sides of the brain.

3

Photographic Interpretation

Thomas Hubbard[1]

3.1 Introduction

I am one happy guy this morning. I am going to talk for about five minutes and then I have arranged for you to talk for the rest of my time. I was worried that it might be like some classes. I open it up for you and then everything dies. I see that will not happen. Each of you is going to get a question that is being passed around. I will explain it as I go along. I will be demonstrating a learning or teaching device. This came about through showing sample photos or slides and getting almost no reaction, or getting the "I like it or I don't like it and I don't know why" reaction with no one going any further.

I am a journalist. When Alyce Kaprow asked which areas you were from, it never got close to me. So, I am not preaching to the converted. I am coming from pretty low-tech, black and white photography. Some of them are displayed in front of you. To get students to understand there is lots of complexity in even the most simple image, I devised the questions that each of you should be receiving. Later, when we get to the slides, I will ask you to use your question.

Students need something to foster learning between elementary understanding and expertise. These questions help before expertise comes along. I have shown this to students and also to photojournalists. I have really enjoyed the reactions of photojournalists and I am looking forward to your reactions.

Photojournalism is in an era that you will be familiar but it is an era of trauma. There is a preoccupation with technology. Just two weeks ago, I looked at about fifty newspapers. I saw lots of good technology. Good technique, good color pictures. But, I did not see

[1]School of Journalism, The Ohio State University, 242 W.18th Avenue, Columbus, OH 43210.

much content. Thats what these slides and photographs will help people to learn.

I talk to editors around the country for an article. Something will be coming out. Their complaint is that the romance is gone from photojournalism. Editors and photographers are saying it. There is a romance and a technique to making a black and white print. Shooting the original image, nursing the print in the developer, massaging it.

Photoshop is traumatic. Digital imaging is traumatic for a photographer who has grown up learning how to interpret reality with a black and white image. Color is traumatic. Digital imaging has finally brought the industrial revolution to the photo department. Before digital imaging, each photographer was an individual craftsperson, able to produce something on his/her own, before others could review it.

A photojournalist is always torn between the unique image seen and the one that is expected by editors. An editor might want a close shot of a speaker at this podium. From the audience, I see another shot. The spotlight above throws a nice triangle on the back curtain here. A publication would not devote the necessary space just to include that unique bit of light behind me. So, the photojournalist might shoot the unique shot, including that triangle, but it would not be used.

All the energy is on the technology. The romance is leaving. I am going to talk for five minutes and then you will get a chance. You will respond to a series of pictures. There are a few of them at the bottom of the stage to give you a preview. With your question, you can respond to the image. They are rhetorical questions. Use them as you will. One drawback to this little exercise is that these are my photographs. I hope you will forget this when you comment. I am here, yes, but I relinquish the privilege of authorship. I have done this several times so I am immune to any of the comments that you come up with. I couldnt find the quote so I am paraphrasing it, once a photograph is viewed by the public, it becomes a public document. It is owned by you as well as the photographer. So please view them in this sense.

The project is Broad and High, an intersection in Columbus, Ohio. Broad and High is the center of the city. It is Times Square, Five Points in Atlanta, it is the very center of the city. It actually is just about the geographic center, certainly the emotional center.

I did a year-long documentary project on Broad and High in 1992. It was not event photography. Photojournalism is often events and exceptions. Thirty thousand cars go down a highway and they are ignored by photojournalists. One car is in an accident and photojournalists rush to that one.

I practice photojournalism and I teach it but I am not completely satisfied with what it is doing. If you eliminate something I characterize as tableaux, that is, rehearsed things, staged things, anything from athletic events to political campaigns to an executive being posed in his or her office; if you eliminate that tableaux, there is not much left in photojournalism. There is a lot more left in reality. Reality with a small "r," because I think there are some philosophers here I dont want to get into an argument with them.

There is drama in the ordinary, which photojournalism often ignores. This is documentary photography. I have come up with my definition of documentary photography.

Documentary is an attempt to photograph a subject with aesthetic enhancement but with as little mediation as possible. "As little mediation as possible" is a goal, it is an ideal, it is not possible by a human being pointing a camera. What I mean by "as little mediation as possible" is trying to go for a goal of a window on reality.

This definition might apply to the output of Fotomat operations around the country. I would think in, say, a thousand years, if an anthropologist could get hold of some photos from this era and had the choice between the Fotomat output and, say, the New York Times photo library, I think they would rather have the Fotomat output.

Realitybased photographs are ripe for interpretation. There is no distraction of events. You deal with, get beyond, or interpret this thing of the objective window. Is it there or is it not there?

Human produced photography can not be random but, in this case, it was attempted. There was no program. The photographer shot what grabbed his attention. Typically, the photographer standing on a corner, glancing at those people over there, that was the picture. It was not a rigid program.

For photographer's information, no contacts were made. The photographer just looked at negatives and put promising ones in the enlarger and made 11 x 14 prints. All this was not a rigid casual, but it was a release from a lifetime of obedient photography. Most photography that you see, publication photography, professional pho-

FIGURE 3.1. Sample figure for interpretation and discussion.

tography, is done by someone at the direction of someone else. This photography was done just by the photographer.

This Broad and High project is work that anyone can comment on. If you have run across the street in the rain, you can comment. If you have stared into the eyes of a traffic cop, you are an expert on these pictures. Certainly, if you are a social scientist or someone who has lived in a city, you can comment. Now, we will see the slides and you can comment. With your questions, you can assume, presume, or speculate. You can even interpolate. So, now the slides (c.f. Figures 3.1-3.4 for sample slides). Does someone have a question that they can apply to this photograph. Maybe as a teacher, I might start pointing to get things started.

3.2 Questions and Answers

QUESTION: I have a question, *Does the technical ability of the photographer add or detract?* I feel that if the photographer has no particular motivation and a great deal of technical ability, chances are he will detract from the image because he will try every trick in the

FIGURE 3.2. Sample figure for interpretation and discussion.

book because there is no mediation of his capability.

QUESTION: What have automatic cameras done, in your opinion. The camera where you just point and it focuses and exposes automatically? How does that affect you.

HUBBARD RESPONSE: I became a believer about four years ago. As I got older and my eyes got a little weaker, I became a believer in the program autofocus camera. Ninety percent of these (photos) were done on pure program. If they did not come out, if the image fooled the program, I had no problem.

QUESTION: There is no correlation to the idea that if you have to write with unerasable ink, vs. a word processor, you are going to think a lot harder about the words you put down on the paper and maybe come out with something better?

HUBBARD RESPONSE: Not because of the technology but these were in effect notes. You are seeing the finals. There may be a hundred rolls of film and there are seventy in the tray. You are seeing less than one per roll. Most photographers cant work that freely. They are required to come back with an image. I went to this location and some days spent three hours there and almost did not take a picture, or did not take a picture that was used. I had that luxury because it was my own project.

QUESTION: I wanted to comment on the image itself rather than talk philosophically in general. I noticed that the image is dark. It is a very ordinary scene but because it is dark and silhouetted it looks very dramatic. I am kind of shocked by the fact that this is a dramatic place that we live in. If you made it lighter it might be boring. I wonder if you would comment on that.

HUBBARD RESPONSE: When I worked at a newspaper, I got the reputation of being the silhouette photographer. This was a chance to get that out again.

QUESTION: *Comment on the credibility of the photo.* It appears that there are two lines that make it credible to me. You have a man crossing the street. There is oncoming traffic. You have the danger of being run over. But, there is a car on the left which is going in a perpendicular direction. There are two lines that define that car, the top and the hood. That makes it credible.

HUBBARD RESPONSE: Thanks.

QUESTION: *What photographic conventions are observed or violated?* I would like to ask you a question. What photographic con-

FIGURE 3.3. Sample figure for interpretation and discussion.

FIGURE 3.4. Sample figure for interpretation and discussion.

ventions are you referring to, aesthetic or technical?

HUBBARD RESPONSE: All. Composition, in this photo, the subject is not centered. There is more space on the left than on the right.

QUESTION: I am so much kinder. Photographic conventions are exactly what I dont want. My question is really a response to your question. What is a photographic convention? Audience comment: Its when a bunch of photographers get together and talk.

HUBBARD RESPONSE: And, they talk about equipment. One convention is composition. Other conventions are, yes, they can be and should be violated. A convention is: you should be in focus on the principle subject.

QUESTION: In this case, it violates that no one is looking at the camera. Not that I necessarily agree or disagree with the value of that.

HUBBARD RESPONSE: Yes, no one is looking at the camera. I will go to another slide and another question.

QUESTION: *How much of this photo is rational and how much is emotional?* This strikes me as being more emotional insofar as we are invited to enter the subjectivity, the mentality of this individual as opposed to any description of this individual, her relationship to society etc. Although there are elements of that here, information about the individual. I think the emphasis is on empathetic understanding.

QUESTION: Why do you like rain so much?

HUBBARD RESPONSE: This is the only photo where I went to an event, the rain. I live west of downtown. The rain was coming. I heard of it on the radio. I rushed down there. It was a gully washer. Think about this, how many weather pictures have you seen in a newspaper. Editors say, "Go out and get a weather picture," like it was just invented.

QUESTION:Three out of the four pictures were of rain.

HUBBARD RESPONSE: Really. This next picture will not be of rain, I hope.

QUESTION: You do like contrast.

HUBBARD RESPONSE: I love contrast. I did this in black and white because I could get all the contrast I wanted. I printed beyond reproduction in some of them. I printed for a showing like this.

QUESTION: Do you think color is necessary for the kind of photographs that you do?

HUBBARD RESPONSE: No. It interferes. I did a year-long project one time in color. It was nowhere as satisfying. With color, the struggle is to get the right color back, to correct it. In black and white, you can interpret.

QUESTION: I have been looking at my question, as you go through it, I start to get it a little more. Mine is, *Is the photo ordered or structured in a true or false way?* I am trying to think along the lines that because of the artists point of view there is a falsity or an alteration to the actual happening being photographed. That presence is very much in each photograph that I am looking at. At the same time, its a matter of degree, Photoshop and how people alter things. That happens even in a simple camera. That presence is there.

HUBBARD RESPONSE: You put me where I am, sometimes I land on one side of that and other times I land on the other. I allow myself to interpret the subject, heightened reality. In other cases, maybe some in this audience do not know of the most famous case of digital manipulation. It was one of the first and most famous cases. National Geographic moved the pyramids. The pyramids were moved closer together to fit on the vertical cover. I abhor that and so does National

Geographic now. They will not move any more pyramids.

QUESTION: My question is, *What rules are followed or broken?* There is an old convention that you do not take pictures into the light. I would like to add another comment. Speaking as an ex-professional photographer. What I have observed in all of your work lined up here. Your comment that you were taking these as unassigned pictures as a personal project is the most telling part of the whole exercise. There is a tremendous literature to the effect that when you are pointing and shooting or using a full scale view camera, what you choose to take is the selection, the selection the photographer makes at the time the picture is taken, that types perception which in turn types affect and cognition and the whole ball of wax at the back of the brain. It is clear to me, as a visual person, that you are delighted by the brightness and interesting shapes. This says a lot about you and your eye and very little about content of what is in the picture.

HUBBARD RESPONSE: Thank you.

QUESTION: As a photographer, there is progression, you grow and develop and change. New equipment and techniques come about. As a photographer are you not interested or curious about developing and growing with the time? You mentioned that you did not like color and do not use it. Maybe that will be a next step with you. We all go through phases. As artists, we grow into and out of phases. All great artists, when you see a retrospective, you see that in their work.

HUBBARD RESPONSE: We live in a world where reproduction is color. Student work, professional newspapers, the movie industry. I like, no, it is my makeup, to be off phase. As the interest in color got higher I turned around. Publications are seeing it also. The very gutsy things you see in publications today, in news magazines, are the black and white essay. Over my lifetime, starting very early when I got a camera, I shot black and white pictures. I tried color when someone told me, "If you ever put color in your camera, you will never go back to black and white." That was a wrong prediction. My absolute joy is black and white. I recognize the other. I am not a Luddite trying to fight it.

QUESTION: *What is strongly in the photo but not explicit?* I get a sense that there is great height here. The window washer is on a cherry picker as opposed to hanging by straps or standing on the ground. The odd angle, its clear that is the top edge, presumably, of a skyscraper or a building. On the other hand, you were real close.

HUBBARD RESPONSE: He was in the air, fairly high. A quick aside here. He came down and asked for a print. This is the bane of professional photographers. I have people all over Cincinnati waiting for that print. Its easier to promise it. I said, "Yeah, Ill send you a print." He really got me."I wash these windows every month. A lot of people take my picture. They all promise to send me a print. Only one person did, a Sports Illustrated photographer really did." That was a challenge. I made a nice 11x14 print and took it to him. A couple months later, I photographed a window washer about ten stories up. From up there, I heard a yell, "I am Ralph! Send me a print!"

QUESTION: *How much did the subject and how much did the photographer contribute?* My initial answer is that the subject contributed the environment and the photographer contributed context and point of view. In looking at some of these photographs, it occurred to me that the photographer can lay in wait for a particular event to occur, like someone jumping over a puddle or a reflection in a window. The photographer can manipulate that environment in ways that are not apparent to the subject.

HUBBARD RESPONSE: Oh, the photographer can lay in wait, yes.

QUESTION: *What is strongly in the photograph but not apparent?* I find that photograph the most effective. There are funny moments in your life where some juxtaposition of yourself with your own image, you recognize yourself. Especially where his arm is. It is totally emotional and not having anything to do with the previous response to that particular question. It had to do with angles and height. The strange face to face "Through the looking glass darkly"' It is very much in that photograph. I think it is emotional.

QUESTION: *Is the composition traditional or daring*? I think its daring in the sense of the painter who painted, instead of a king, he painted contemporaries. All these pictures have a very ordinary person who is raised to a wonderful level drama and dignity. We know he is poised on the cherry picker, poised in a moment of time, like a really important person.

HUBBARD RESPONSE: There was a New York photographer who went by the name Weegee. He was the subject of a movie about a year ago. He said something I have always tried to follow. "Never rob anyone of his song." With a camera, you can do a number on someone. Give me a camera and give me twenty minutes with you and I can do a number on you. I do not rob people of their songs. Thank you.

QUESTION: Commenting on the last question, we have to remember that this is one out of thirty-six. You took a lot of pictures to get these. I see prints, out of a darkroom. You held your hand over the light and all that stuff. You twiddled the chemicals and got what you wanted. What are we looking at on the screen, a picture of a picture or a negative, what is this?

HUBBARD RESPONSE: It is a Kodachrome copy of the print.

QUESTION: The dark sky, is that an orange filter?

HUBBARD RESPONSE: It is not the sky, its part of the building. It is representative of about 20 prints in the trash can before I got it.

QUESTION: I have a question that sort of been irritating me. I am not sure how to interpret it. Ill just ask it. *Does this photo teach?* I am not sure how to react to "teach." What does that mean?

HUBBARD RESPONSE: You answered it with your question. The questions came from, partly dreaming them up, partly from reading reviews of art and photography. I would turn a statement into a question. That is probably where the question came from. Does it teach? A man built a wooden structure on the Ohio State University campus. He was an artist. People wondered what it was. I wondered what it was. It did not look like art. No one knew it was art. They thought it was a hamburger stand. I finally decided, after thinking about this for three years, that was his point. To make you wonder what it was. So maybe that is teaching.

QUESTION: There was a tremendous controversy about the Maplethorp exhibit in Cincinnati. I wonder whether you have any thoughts about that particular controversy and any reaction to the type of work that you do and the type of work he did. You both work in black and white.

HUBBARD RESPONSE: There is a fairly simple answer. That controversy came about because of a prosecutor who can get headlines being a puritan. He pursued Hustler magazine for pornography. Interestingly, all the media persons in Cincinnati realized how dangerous that was. If he could stop Hustler magazine, tomorrow, he could stop something else. He is still in business. People walk around town saying, “Cincinnati is a conservative town.” It is really not,it is a fairly modern town. I think seventy five percent of that Maplethorp thing was the district attorney.

I promised Professor Marchese I would not run over. Thank you very much.

4

Cognitive Origins of Graphic Productions

Barbara Tversky[1]

4.1 Introduction

Long before there was written language, there were pictures. Maps, whether drawn in sand or on paper, whether inscribed in stone or carved from wood, appeared in ancient cultures all over the world. Ancient cave paintings and petroglyphs depicted animals and ancient pottery tokens and notches in bones represented accounts. These surviving pictorial relics are probably but a small fraction of the wide use depictions must have had. Although petroglyphs and stelae and cave paintings and pottery shards survive, no interpreters of their inscriptions do, so we can only speculate about their meanings. The remarkable compendium of Colonel Garrick Mallery, *Picture Writing of the American Indians* (1893/1972) [1], gives us a contemporaneous glimpse into the many functions of pictorial language. In 1876, while stationed in the upper Midwest in the military, a pictographic calendar of the Dakota nation came into his possession. He published it with interpretation, attracting the interest of the Secretary of the Interior, who requested Mallery's services. The Secretary of War obliged by ordering Mallery to continue his field work in ethnology. Mallery went on to gather a valuable collection of hundreds of examples of picture writing, including calendars, histories, legends, sayings, stories, and letters. He was able to verify the interpretations of many of them by consulting the people creating and using the picture writing just as this form of expressing thought was being replaced. Two examples appear in Figure 4.1. The first is a birch bark notice informing people that the writer had gone across the lake to hunt deer. The second is also on birch bark, and depicts a

[1]Department of Psychology, Bldg. 420, Stanford University, Stanford, CA 94305-2130.

battle between the Ojibwa and the Sioux in which the Ojibwa lost one man.

How can these depictions be characterized? Of course, generalizations are problematic. Many of the more sophisticated communicative depictions consist of discrete elements in a linear array. The elements appear to represent persons or objects or events, and the linear array seems to correspond to a sequence. The sequence is typically temporal, either the sequence of events, as in the second example (where time goes from right to left), or a sequence of thoughts/words, or both, as in the first example. Simpler depictions, like many petroglyphs, often consist only of icon-like elements, with no relations between them.

FIGURE 4.1. Pictographic messages on birch bark (Ref. 1). A) "I am going across the lake to hunt deer" (p. 331); B) Battle between Ojibwa and Sioux (p. 559).

Many of the surviving depictions seem to have been used in communication and record-keeping. Other depictions seem to have had aesthetic and sacred functions. Depictions have served all the functions of written language, and then some. Depictions are compelling. Compared to purely symbolic script, they are easy to produce and easy to understand. Indeed, in his first encounter with pictures at an early age, a child was able to correctly label simple line drawings of objects he was familiar with from real life [2]. Pictures are increasingly used on highways, in airports, in books, in the mass media, and in computers. Depictions are not just used, their current popularity and ease of production has insured that they are also abused(for examples, see Tufte [3], Wainer [4]). Of course, interpreting depictions is not always immediate, and often depends on shared conventions. Even if we recognize an icon as an airplane, it is not immediately obvious that it stands for an airport, though it is undoubtedly more readily interpreted than a comparable word in an unknown alphabet.

The graphic inventions for both concrete and abstract concepts produced by different cultures show remarkable similarities, and show similarities to the graphic inventions produced by children. The changes in graphic inventions over time within cultures often parallel the changes over time within children. One theme of this paper, then, is that ontogeny recapitulates phylogeny in the production of graphic inventions. Some of the uses of space to convey meaning in depictions have parallels in language and gesture as well. These similarities and parallels suggest that the depictions of many concepts and relations are natural or cognitively appealing, that there are cognitive principles underlying the similarities. That is the second theme of this paper. All along, I will discuss a number of graphic inventions produced by children, and compare them to historical examples for similar concepts. I will attempt to draw some cognitive principles for graphics from these examples.

4.1.1 Elemental and Relational Concepts

It is common in discussions of language to distinguish objects or elements from attributes or relations or predicates, the subjects of discourse from the qualities or activities attributed to them. This distinction has parallels in the grammatical distinction between subject and predicate. It also has parallels in the distinction made in cognition between schemas or data and operations performed on them. This is a functional distinction, and while it holds for many cases, it breaks down for others. Two assertions about elements and relations seem self-evident. First, it is easier to depict concrete elements than to depict abstract elements or to depict relations, concrete or abstract. Depicting items of food is easier than depicting a restaurant, which, in turn, is easier than depicting a law firm. Depicting a customer is easier than depicting making a purchase and depicting a car is easier than depicting renting a car or selling one. The second assertion is related. There seems to be a bias to interpret iconic depictions as things, outcomes, or states, rather than as ongoing activities, changes, or processes. A depiction of a plow is more likely to be taken to represent the object plough than the act of plowing, and a depiction of a leg is more likely to be taken as the object leg than the act of walking, though in early Sumerian, as can be seen from Figure 4.2, a plow-like logograph stood for "to plow" and a leg-like logograph stood for "to go" [5].

4.1.2 How Depictions are used to Convey Meaning

There seem to be two basic ways that depictions convey meaning. The first is through pictographs or symbols, that is, drawn entities that are meant to stand for concepts directly or for elements of language that refer to concepts. The other way that depictions carry meaning is by the spatial arrangement of the pictographs or symbols. For the most part, pictographs or symbols have been used to represent elements, and pictorial space to convey relations between elements, but there are notable exceptions, including the arithmetic operations to be discussed shortly.

Now a word about terminology. Instead of the rather cumbersome term "graphic invention," I will often use the term "representation." Here, "representation" will refer to something drawn on paper or displayed on a computer screen, or inscribed in clay, something out there that everyone can see, rather than the typical sense of representation in cognitive psychology as a private mental structure that is presumed to intervene between knowledge out there and mental processing.

4.2 Pictographs and Symbols

4.2.1 General Principles

4.2.1.1 Icons and "Figures of Depiction"

To represent elemental concepts, people have long used icons. This is straightforward when the icon is a depiction of the thing to be represented, such as a fish or a bird or a horse. Often a depiction of the entire object is used to represent the object, but in other cases, perhaps for simplicity, a part of the object, usually a significant part,

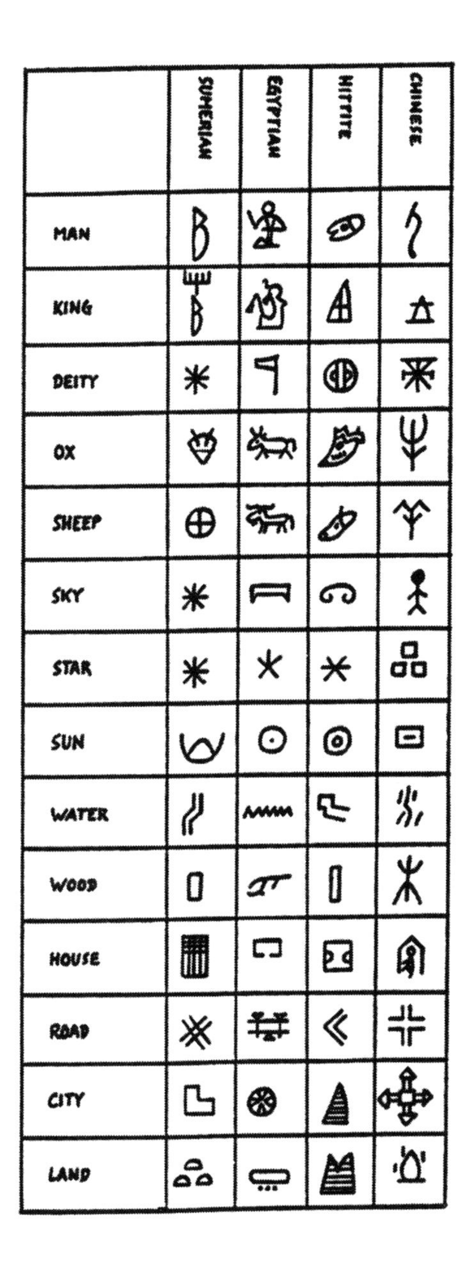

FIGURE 4.2. Pictoral signs in the Sumerian, Egyptian, Hittite, and Chinese writings (Ref. 6 ,p. 98). Reprinted with permission.

is used to represent the object. In Figure 4.2, the Hittite sign for a man was the head of a man, and the Sumerian sign for an ox was the head of an ox. The Crow Indians used horse foot prints to signify and quantify horses [1]. These are examples of synecdoche, using a part to represent a whole. Synecdoche is common not only in pictorial signs, but also in figures of speech, using the same principle. Ranchers count "head" of cattle, and teenagers drive their "wheels" to school. Related to synecdoche is metonomy, where an associated object is used to represent an object. To convey famine, the Dakotas portrayed empty racks for drying buffalo meat [1]. In Figure 4.2, all of the signs for water use the movement of water to portray water. Crowns are used to represent kings, both in pictures and in speech. "The church" and "the White House" may refer to buildings, but they may also refer to the institutions the buildings symbolize.

4.2.1.2 Rebus Principle

Indeed, most authorities agree that most writing systems began as pictures, presumably representing meaning directly [6]. Representing meaning directly quickly becomes problematic. For example, it is difficult to represent abstract concepts and proper names. All true writing systems solved those problems by inventing ways to represent sound. One widespread means of representing sound is based on the rebus principle where a pictograph of an object with a similar name is used to represent something that cannot be easily depicted [5, 7, 8]. An example would be writing "before" using a picture of a bee and the numeral 4. The contemporary use of hearts to mean "love" in bumper stickers is an application of the rebus principle. There is recent and still controversial evidence that Sumerian cuneiform, the world's earliest writing system, grew out of the shapes of clay tokens used in accounting, rather than from depictions [9]. Nevertheless, the evolution of cuneiform from representing meaning directly to representing sound presumably followed the same rebus principle.

4.2.1.3 Schematization

Even when an icon is used in a straightforward manner to represent the thing the icon depicts, icons are schematic and undergo further schematization over time. That is, icons are simplified representations of classes rather than representations of classes. The reasons for this are many. With constant repetition, ease of execution becomes as important as ease of recognition. Schematic icons take less artistic talent and less time to produce. Artistic detail is not essential for effective communication. Further schematization is usually a joint product of efficiency and the representing medium. Cuneiform was inscribed using a stylus in wet clay, whereas hieroglyphs were drawn on papyrus (or inscribed in stone). Although many cuneiform patterns and glyphs started iconic, they became schematized, and in different ways (Figure 4.3). For cuneiforms, dramatic changes came when larger clay tablets were introduced. In order to hold the tablets comfortably, the signs were rotated ninty degrees, leading to a loss of iconicity. Signs that were drawn or painted underwent other changes. Because modern icons, such as those used on highways or in airports, are reproduced mechanically rather than by hand, the pressures toward schematization are less.

4.2.1.4 Conventionalization

Many concepts are difficult to represent pictorially, even applying "figures of depiction." Depictions can be based on acoustic associations as well as visual associations, as in the applications of the rebus principle. Depictions may also be totally arbitrary. In all of these cases, and in the cases where pictographs become so schematic that their depictive origins are obscured, when the symbols become accepted by a group of users, they become conventionalized. Conventionalization is a normal process occurring in all forms of communication.

4.2.2 Children's Early Writing

Ferreiro and Teberosky [10, 11, 12] working in Latin America and Tolchinsky-Landsman and Levin [13, 14, 15] working in Israel asked preschool, preliterate children to write words or sentences. This is not a pure task in inventing representations because these children were growing up in environments with books and other printed matter, so they had exposure to writing, and even perhaps to the notion that writing represents sound. Most of the early inventions of children were uninterpretable, but did reveal a number of characteristics of writing.

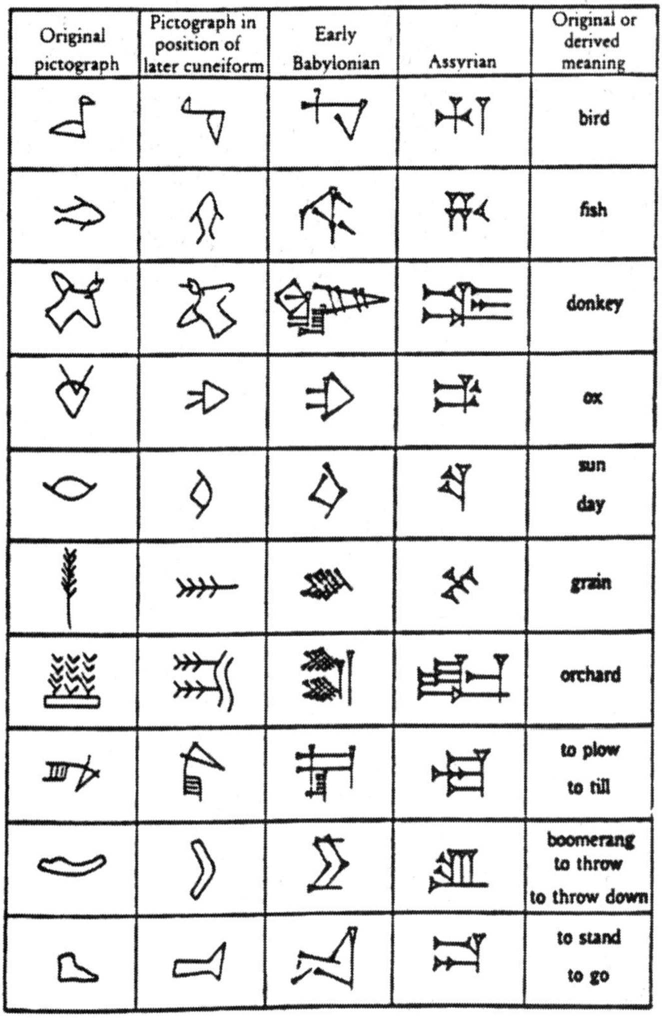

Original pictograph	Pictograph in position of later cuneiform	Early Babylonian	Assyrian	Original or derived meaning
				bird
				fish
				donkey
				ox
				sun day
				grain
				orchard
				to plow to till
				boomerang to throw to throw down
				to stand to go

FIGURE 4.3. Rotation and loss of iconicity of some Sumerian signs. (Ref. 5, p. 74). Reprinted with permission.

4.2.2.1 Spatial Arrangement

For one thing, the overall organization of children's early writing was linear, as is true of all writing systems in the world. Some writing systems have a columnar organization, some have a row organization, and some, like Japanese, have both, but all systems have a linear organization, perhaps reflecting linearity of speech. In these experiments, sentences were short and simple, and separate words were enunciated as such. Children tended to produce one mark, sometimes a complex mark, for each word. That is, they discerned words as units and used empty space to separate them. Although we take the concept of a word for granted, that concept seems to have arrived together with the advent of written language (small, personal communication). And although we assume that written languages separate words spatially, that hasn't always been the case. Classical Greek was written continuously, with no breaks for words, sentences or paragraphs.

4.2.2.2 Symbols

Children produced marks for objects (nouns) earlier than they produced marks for activities (verbs). Since many of the marks children produced for words were not iconic, this suggests that the priority of signs for objects over actions is more than just the greater accessibility of icons for objects than for activities. Let's look more closely at the characteristics of the marks children produced for objects. Many of them resembled the objects in one way or another. For example, if given a choice of colored markers, children often chose a marker of the same color as the object, say, red for apple and tomato. Often, children's marks matched the objects in shape, say, round for ball and long for rope. Color and shape are strong influences on young children's categorization (cf. [16]), and shape is a strong influence on young children's use and generalization of their early words [17]. Size of the object, too, was reflected in children's marks for words; elephant got a larger mark than ant. Yet, although many of the children's marks resembled the objects in physical characteristics, others reflected properties of the sounds of names of the objects. For example, objects with longer names got longer marks, and repeated sounds got the same symbols.

4.2.2.3 In Sum

The attempts of preliterate children to write reveal many of the characteristics of writing systems developed all over the world. They display a linear organization, they separate words, and they produce marks for concrete objects prior to marks for activities. In producing marks for words, children used the same correspondences that written languages have used, correspondences of appearances of the objects and the written signs, that is, representing meaning directly, and correspondences of the sounds of the names of the objects and the written signs, that is, representing meaning by representing the sound system of language.

4.2.3 Children's Early Arithmetic

Unlike fully developed writing, arithmetic symbols do not reflect the sounds of the names of the numerals or operators. Mathematical writing is a visual system, not a phonetic system. In truth, many modern day writing systems have visual, non-phonetic elements as well. The visual elements of writing include spacing between words, punctuation, numerals, and certain signs important enough to be included on keyboards, %, $, and &. Although mathematical writing does not reflect sound, current mathematical writing does not reflect appearance either, though, like writing, historically, some aspects of it did.

4.2.3.1 Representing Numbers

Hughes [18] asked preschool and early school-aged children to represent various arithmetic concepts on paper. In one task, he put one to six bricks on a table, and gave a child a piece of paper, asking the child, "Can you put something on paper to show how many bricks are on the table?" (p. 55). There were four types of responses. *Idiosyncratic* responses were squiggles, bearing no relationship to the number of bricks. *Iconic* responses were pictures of things other than bricks, where the number of things corresponded to the number of bricks. Often the iconic responses were lines, like tallies, but sometimes they were unrelated objects, like houses. These two categories of responses were dominant in the preschool years, but dropped to very low levels in the first grade. *Pictographic* responses were pictures of bricks, where the number in the picture corresponded to the number on the table. This type of response increased in frequency from preschool to first to second grade, and dropped in third grade. Finally, *symbolic* responses used the conventional numerals. This type of response was infrequent in preschoolers, but increased steadily thereafter, so that by third grade, it was the dominant response.

4.2.3.2 Representing Zero

Hughes' next step was to brush aside the bricks, and to ask the child to "show that there are no bricks on the table" (p. 63). The children who were using the numeric symbols put down zero with little difficulty. Children who did not know the numeric symbols, however, were often confused, and did not know what to put. Some left a blank, and some drew a line. In short, amongst children who did not know zero, there was considerable confusion and little agreement.

4.2.3.3 Representing Addition and Subtraction

Hughes and his collaborator Jones [18] used techniques similar to those that extracted number representations from children to extract representations of addition and subtraction. In one task, they put two bricks on the table, and then added two more, and asked, "Can you show that we first had two bricks and then added two more?" (p. 72). In another task, they removed one brick from a pile, and asked, "Can you show that I took one brick away?" (p. 73). These questions stumped the children. Many simply recorded the total number of bricks, or the numbers of the two sets of bricks to be added. None of the children, not even those who used arithmetic daily in the classroom, used the standard plus and minus signs

Some of the children drew hands to show the action of adding or subtracting, or feet to show bricks walking away. Other children used arrows. On the whole, the depictions were not satisfactory representations of these abstract concepts.

4.2.4 Historical Examples of Arithmetic

4.2.4.1 Counting Systems

The children's inventions were then compared to the history of written number systems in other cultures in support of the idea that ontogeny recapitulates phylogeny. Except where noted, the historical discussion comes from Hughes [18]. Prior to recording numbers, whether by writing or notching, it is likely that people used parts of their bodies, especially fingers, to count and to represent sums. Systems for keeping track of quantities using fingers and other body parts appear in many cultures. Words for numbers are often related to words for fingers, take our own "digit," for example.

4.2.4.2 Representing Numbers: Pictographs

The first known system for representing numbers was (by Hughes' definition) pictographic [9]. In ancient Sumeria, a single ovoid (jar-like) token stood for a single jar of oil and five ovoid tokens stood for five jars of oil. Other items were counted by tokens with other shapes. Later, the tokens were replaced by incisions in clay, but the principle of counting different types of things with different types of symbols remained. Schmandt-Besserat [9] argued that this way of keeping records of property and property transactions drove the development of written language.

The principle of counting different kinds of things with different kinds of symbols is reflected in language in classifiers. Although classifiers are not common in English, there are some common examples. We don't request "five papers" (unless we mean scholarly papers or newspapers); instead, we request "five sheets of paper." Similarly, we refer to "six slices of bread" and "three sticks of gum." Classifiers are mandatory in many languages, where use of specific classifiers typically depends on object shape, much like "sheet," "slice," "stick," and "roll" in English. Modern-day analogs of Sumerian pictographic counting are to be found in Isotypes, a system for pictorial graphing invented by Otto Neurath [19], in which bar graphs are constructed from icons representing the quantified elements, such as using sheaves of wheat, barrels of apples, and pairs of shoes to indicate yearly output for different countries (note that sheaves, barrels, and pairs serve as classifiers in the previous sentence).

4.2.4.3 Representing Numbers: Tallies and Symbols

After pictographs came tallies, which appeared with great frequency in ancient systems. In Sumeria and probably elsewhere, tallies preceded written language. Like pictographs, tallies preserve a one-to-one correspondence between the number of objects and the number of written marks. Tallies are more abstract than pictographs because they do not reflect the kind of thing being counted. Since what is important for counting is the number and not the object, a neutral mark that is easy to make suffices, exactly what was produced spontaneously by many cultures and by preschool children. Tallies, however, become cumbersome for large numbers, and are difficult to use in operations on numbers. This creates pressure to develop symbols for the numbers. Number symbols evolved in ancient Egyptian. The Semitic languages used letters to represent numbers, a system borrowed by the Greeks. This system was compact compared to tallies, but did not have a place holder, so that the relations between multiples of ten were obscured.

Readers may notice that in Hughes' young subjects, tallies were replaced by pictographs, whereas in Sumerian history, tallies replaced pictographs. As was noted, tallies are more abstract than pictographs as they can be used for counting anything, whereas pictographs can be used for counting only one kind of thing. Hughes' situation did not encourage the development of abstraction as he only had the children representing numbers of bricks. If he had requested that the children represent the quantity of a number of different kinds of things, tallies might have replaced pictographs, as they did historically.

4.2.4.4 Representing Zero

Symbols for zero evolved far after symbols for numerals, and many widely used systems did not have a zero. Some ancient number systems, such as the Babylonian and the Maya, did develop ways to denote zero, both as the empty set and as a place holder. The Babylonian system did not survive in western mathematics. The present day zero (like the present day number symbols) arrived to the west from India via Arab mathematicians. It was adopted to represent the empty set in the seventh century, and as a place holder about two hundred years later.

4.2.4.5 Representing Addition and Subtraction

Historically, special symbols for arithmetic operations were a late development. The Egyptians used pairs of walking legs, from left to right to signify addition, and from right to left to signify subtraction. An Alexandrian mathematician used an upward pointing arrow to denote subtraction. Both these symbols imply motion, and both were used by children in Hughes' studies for similar purposes. None of these symbols was widely used, and none survived. The plus and minus signs in use today appeared in Germany in the fifteenth century, and the equal sign about a century later.

4.2.5 Historical and Developmental Parallels in Arithmetic Representations

The parallels between children's inventions and historical inventions of numbers are striking. Prior to recording, body parts, especially fingers, were and remain used for counting and keeping track of sums. Early permanent records, Sumerian tokens, counted different kinds of things with different tokens, a practice that endured after writing began. For ancient Sumerians and Hughes' young children, the number of sheep or bricks was indicated by a one-to-one correspondence between items in the world and pictures or tokens of the items. Later, tallies replaced pictographs or tokens. In tallies, the same marks are used to represent quantities irrespective of the type of quantity. Still later, number symbols replaced tallies. Number symbols are more efficient, both in conserving space and in arithmetic manipulations. The children most likely did not invent number symbols, rather, they seemed to be applying what they were learning in school. Historically, number symbols were slow to be developed, and symbols for arithmetic operations appeared even later. Complex as elemental concepts can be, relational concepts can be even more complex, and consequently more difficult to depict.

4.3 Pictorial Space and Pictorial Devices

4.3.1 Children's Graphic Productions

The arithmetic operations have generally been represented by symbols. Another way to represent relational concepts is to use spatial relations in pictorial space. Maps are a simple and straightforward example, where distances and spatial relations among cities are represented in miniature by distances and spatial relations on paper. Another prevalent modern example are graphs. In contrast to maps, graphs are a relatively recent invention. Graphs began to appear in significant numbers in Europe only in the late eighteenth century [3]. I turn now to discuss cross-cultural research on children's graphic productions for a variety of concepts.

4.3.1.1 Task

Kugelmass, Winter, and I [20] were interested in how children use space to represent increases in abstract concepts, temporal, quantitative, and preference. The task we developed was simple. We sat next to a child and gave the child a square piece of paper and some stickers. We told the child that we were going to put down a sticker for, say, breakfast time, and the child should put down one sticker for lunch time and another sticker for dinner. That was one of the temporal concepts, and there were one or two others depending on the experiment. Examples of a quantitative dimension were the number of books in a child's backpack, the number of books at home, and the number of books in the library. For preference, we asked the child for their favorite food, a food they dislike very much and a food they sort of like. For all these attributes, the task was the same. The experimenter put down the first sticker in the middle of the square page, and the child put down the next two stickers.

4.3.1.2 Information Preserved

We were primarily interested in three aspects of the data. The first was *information preserved* in the children's mappings on paper. Would they see the relation as *nominal*, as three unrelated items, and put the stickers down in a disorganized fashion, or would they see the relation as *ordinal*, and put the stickers on a line? In the second and third studies, we asked children about inequally spaced items, giving them the opportunity to preserve *interval* information in their mappings. In nominal representations, things belonging to the same category are placed together, and there are no relations between categories. In ordinal representations, the order of the items is meaningful and interpretable, and in interval representations, the intervals between items are meaningful and interpretable, the greater the spatial distance between items, the greater the conceptual distance. As expected, there were age effects in the information preserved in mappings. Some of the kindergartners and first graders preserved only nominal relations, though most of them preserved ordinal information. Only a few of the fifth graders preserved interval information, though with considerable coaching, some of the third graders came to represent interval. There were some effects of content as well. Interval information was represented at an earlier age for the most concrete concept, time, next for the quantity, and latest for the most abstract of the concepts, preference.

4.3.1.3 Directionality

The second aspect of the data that interested us was directionality. When children's mappings preserved order or interval, what direction would increases go? We had two thoughts on the issue. One obvious influence on directionality would be direction of writing. Sonny and his collaborators had found effects of writing direction on a variety of tasks [21, 22, 23]. In these studies, we ran large numbers of English-speaking children in the U. S., and Hebrew-speaking and Arabic-speaking children in Israel. Because of writing habits, we expected that English speakers would have strong left-right directionality, and Arabic speakers strong right-left directionality, with Hebrew speakers in between. There are several reasons why right to left directionality would be stronger in Arabic speakers than in Hebrew speakers. First, Arabic script is connected and each character is written from right to left, whereas Hebrew is disconnected, and most characters are written left to right [24]. In Hebrew, arithmetic is written right to left, as in the West. Arabic-speaking Israelis first learn numbers left to right, and switch to the Western pattern in the middle of schooling. Finally, Hebrew-speaking Israeli children are more likely to have contact with European languages than Arabic speaking Israeli children.

Another influence on directionality might come from cognitive correspondences between direction and quantity. These correspondences are readily revealed in language, where "up" is on the whole associated with more and good, that is, with positive value, and "down" with less and bad, that is with negative value [25, 26, 27]. We say, "I'm feeling up today," "She's at the top of the heap," and "His standing dropped." "High society" contrasts with the "underworld." Comparable expressions appear in many languages, suggesting that cognitive universals underlie the correspondence. From this, we would expect that quantitative, especially evaluative, relations would be mapped so that increases were vertical, from low to high.

The data corroborated both intuitions. Writing direction was reflected in the mappings for temporal concepts, but not for any of the other concepts. In English speakers, time increases were mapped with greatest frequency from left to right; in Arabic speakers, time was most frequently mapped right to left; and Hebrew speakers were in between, primarily using the horizontal dimension and both directions about equally. For quantity and preference, there were no effects of writing direction. For both concepts and all cultures, the direction of increases split into approximately equal thirds, going left to right, right to left, and down to up. The only direction that was avoided was the one that is incompatible with the cognitive correspondence, mapping increases downward. These effects did not change with age, so that the proportions of use were about the same in high school and college aged participants as in the younger children. Neither children nor adults seemed to be borrowing their mappings from conventional graphs despite the fact that by fifth grade, children have begun studying graphing. By convention, increases in graphs go from left to right or from down to up. Even the young adults did not follow convention; instead, they seemed to be following spontaneous inventions.

4.3.1.4 Content Dependence

From the previous results, it is clear that children and adults from three language cultures map different concepts differently. The graphic inventions were not content independent. In this situation, children do not have a general graphing schema for increases that they then apply to any concept, regardless of content, the way that tallies can be used to count any kind of quantity. It is possible, of course, that with more experience, people's graphing inventions would become conventionalized as in graphing practice.

4.3.1.5 In Sum

Children and adults use space on paper readily and systematically to express increases in a variety of concepts. For temporal concepts, the direction of increases corresponded to the direction of writing. Since dates and times of events are often incorporated into writing and since calendars and schedules correspond to writing direction, there is a close association between writing and time, closer than between writing and the other concepts. Small [28] examined Etruscan and Greek vases for the directionality of the scenes depicted on them. The scenes tell stories that occur over time, often the same stories, but the direction corresponds for the most part to the direction of writing, right to left in the case of Etruscan and left to right in the case of Greek.

For quantitative and preference concepts, there was no single dominant direction for increases. Instead, upwards and the two horizontal directions were used about equally often. The downwards direction was never used for expressing increases. The upward direction corresponds to pervasive expressions in language associating up with good or more. This suggests that there is a natural cognitive correspondence between the upward direction in space and positivity that is revealed in both language and graphic productions (cf.[25, 26, 27]).

4.3.2 Spatial Pictorial Devices

Now, I'd like to apply these findings along with examples from the "visual language" we have all around us to point out some general principles for the way spatial relations and related spatial devices are used to convey meaning, primarily to represent relations. Many of the relations are based on Gestalt principles, especially those of grouping, in particular, proximity, common fate, and similarity. The device of primary interest is the use of spatial relations to convey other relations. Other pictorial devices are also commonly used, such as size, color, and highlighting of elements. The devices separate neatly into those conveying nominal or categorical relations, those conveying ordinal relations, and those conveying interval relations.

4.3.2.1 Categorical Relations

The simplest devices are those used to group elements into classes, sharing a single feature or set of features. One device has already been mentioned, separating the letters of one word from the letters of another word by leaving a space between words. Indentation of paragraphs is another example, where major ideas are separated by a spatial device. Empty space is not the only device for delineating groups. Another common practice is delineation, using a line to enclose a group, as in a frame, or parentheses, or a box, or a circle. Color, shading, and cross-hatching are other ways to signify that some elements are related by being in the same category.

Some spatial devices group and juxtapose at the same time. A good example is organization into rows and columns, where, again, space separates items along the rows and the columns. Usually, the column items are related to each other in one way and the row items are related to each other in another way. Rows and columns cross-classify items into several groups at the same time. Think, for examples, of train schedules, or of demographic tables, such as one of countries, with their basic statistics, population, area, GNP, and so forth. Lines or dots can also be used in addition to empty spaces to cross-classify.

Another well-developed example of using depictions for conveying categorical relations is Venn or Euler diagrams (cf. [29, 30]). In essence, Venn diagrams are based on object perception. Displays of disjunction show separate, non-contiguous objects, displays of intersection show objects overlapping, and displays of inclusion show one object enclosing another. Thus the physical relations between objects are extended metaphorically to represent the logical relations among sets of elements.

4.3.2.2 Ordinal Relations

Ordinal relations can vary from a partial order, where one or more element has priority over others, to a complete order, where all the elements are ordered with respect to some attribute or set of attributes. Printing the elements in the order of the relevant attribute, say children by birth order or students by GPA, is a simple spatial device for signifying order. When elements are written from first to last, as is common, then the order of increase is opposite the order of writing, a pattern found in the work reported on graphic productions in children for concepts other than time. Another simple spatial device to convey order is the use of indentation in lists and in outlines. For complete orders of a complex set of elements, trees and graphs are useful. Expressing order is one of the many uses arrows serve, for example, the sequence of operations. Relative size, superposition, bullets, and highlighting are some of the other pictorial devices that are used to signify order.

4.3.2.3 Ordinal Relations: Directionality

From the psycholinguistic literature discussed earlier, it appears that the horizontal dimension is neutral, but the vertical dimension is not. For the vertical dimension, upwards is associated with good and more. Spatial expressions like "I'm sitting on top of the world" and "he's sunk into a deep depression" seem to occur in languages all over the world. The association of up with positivity occurs not just in language but in gesture as well; thumbs up and high five are success, thumbs down is failure. The only well-known exception to this rule is inflation, though increased inflation isn't bad for everyone. Physically, being "on top of" means "controlling." Thus the association of up to more and better and stronger is not just perceptual, but functional as well. Musical notation also follows this convention, where there seems to be a natural mapping between musical pitch and height on a score. The horizontal dimension is more neutral; both those on the left and those on the right seem pleased with their positions. Increases went either direction in spontaneous graphs of quantitative and preference concepts in children and adults from different writing cultures. In conventional graphing, increases go from left to right, corresponding to the order in which Western languages are written (graphing was a Western invention), or from down to up, corresponding to the cognitive universal of up with positivity. Most graphs appearing in newspapers, journals, and textbooks have time as one component, and time, a neutral dimension, is conventionally plotted horizontally.

4.3.2.4 Ordinal Relations: Directionality of Scientific Charts

When we think about evolution, we think about the evolutionary tree. It has become an icon of that branch of biology. Several other scientific disciplines have graphic displays closely associated to them, the language tree in linguistics, and the geological ages in geology. We sampled all the textbooks in biology, linguistics, and geology that we could find in the Stanford Library that had these general graphic displays to examine their directionality. In seventeen out of eighteen of the evolutionary charts, human beings are at the top. In fifteen out of sixteen of the geological charts, the present was at the top. In thirteen out of fourteen of the linguistic trees, the proto-language was at the top. In contrast to the vast numbers of graphs that plot time horizontally, in these charts, time runs vertically. The directionality of time cannot explain the pattern, however, as the present time is at the top for the evolutionary and geological charts, but at the bottom for the linguistic charts. What the charts have in common is the presence of some sort of ideal at the top, whether the ideal is man as opposed to the beasts, the present time as opposed to the past, or a proto-language as opposed to its' variegated offshoots.

One other aspect of these scientific charts and trees, and others like them, is of interest. When the trees are vertically arrayed, as in the overwhelming number of examples, the vertical relations are significant. They cannot be changed without altering the meaning of the chart. The horizontal relations are arbitrary; they can be permuted without any change of meaning.

4.3.2.5 Interval Relations

The use of space to represent interval (or ratio) relations is the simplest of all. The degree of proximity in space reflects the degree of proximity of the relation being graphed, time, income, preference, whatever.

4.4 Conclusions

4.4.1 Ontogeny Recapitulates Phylogeny

We have reviewed a medley of investigations of conveying meaning pictorially. Some of that research was on children: their early attempts at writing words, their early representations of arithmetic, and their early graphic productions. Some of the research was on cultural inventions: communication of ideas through pictographs, writing, maps, and graphs. There are large and fascinating areas we have not reviewed, other work on maps and writing, and work on color, musical notation, dance notation, and the like. Enough has been reviewed to support the opening claim, that the graphic representations produced by children and historically bear striking similarities, and change in similar ways.

4.4.2 Two Aspects of Meaning and Two Aspects of Representations

In analyzing the graphic inventions of children and cultures, we made two distinctions, one about what aspect of meaning was being represented, and the other about what aspects of depictions were being used for the representation. For aspects of meaning, we distinguished representing objects or elements from representing relations between objects. If availability of depiction is a measure, then it seems that it is easier to represent elements than relations, and easier to represent concrete elements than abstract ones. Representations for concrete elements appear earlier, in children and in cultures, than representations for abstract elements, and representations for elements appear earlier than those for relations. For aspects of depicting, we distinguished pictographs or symbols from pictorial devices, especially spatial relations, but also emphasis devices like size and highlighting. In many cases, pictographs and symbols are used to represent elements, and spatial relations are used to represent relations between elements. This seems natural. However, there are many relations that are represented by symbols, perhaps because they are very common, perhaps because they are refined.

4.4.3 Cognitive Naturalness

The selection of pictographs and pictorial devices is principled. Wherever possible, elements are represented by depictions of them. When elements are abstract or otherwise difficult to depict, then they are often represented by related objects, parts of objects, or symbolic objects. These are the same devices used in literature, metaphor, metonomy, synecdoche. Justice is represented by scales in depictions as well as in speech. As for relations, elements that are closely related are depicted closer than elements that are distantly related. Spatial proximity serves as a metaphor for abstract proximity. As for elements, the underlying spatial metaphor is revealed in figures of speech (and in gesture) as well as in depictions. We can "feel close to" people we've never met, but who share our values. The pervasiveness and systematicity of these depictive devices for expressing abstract meanings suggest that they are cognitively natural.

4.5 Acknowledgements

I am indebted to Jocelyn Penny Small and Rochel Gelman for insightful discussion and invaluable references. Preparation of this manuscript was aided by the Air Force Office of Scientific Research, Air Force Systems Command, USAF, under grant or cooperative agreement number, AFOSR 89-0076. Some of the research was supported by NSF-IST Grant 8403273 to Stanford University and by grants from the Human Development Institute and the Cognitive Psychology Institute of the Hebrew University.

4.6 References

[1] Mallery, G. (1893/1972). *Picture Writing of the American Indians.* (Originally published by Government Printing Office). NY: Dover.

[2] Hochberg, J. & Brooks, V. (1962). Pictorial recognition as an unlearned ability: A study of one child's performance. *American Journal of Psychology*, 75, 624-628.

[3] Tufte, E. R. (1983). *The Visual Display of Quantitative Information.* Cheshire, CT: Graphics Press.

[4] Wainer, H. (1980). Making newspaper graphs fit to print. In P. A. Kolers, M. E. Wrolstad and H. Bouma (Editors), *Processing of Visible Language* 2, pp. 125-142. New York: Plenum Press.

[5] Coulmas, F. (1989). *The Writing Systems of the World.* Oxford: Basil Blackwell.

[6] Gelb, I. (1963). *A Study of Writing.* Second edition. Chicago: University of Chicago Press.

[7] Coe, M. D. (1992). *Breaking the Maya Code.* NY: Thames and Hudson.

[8] DeFrances, J. (1989). *Visible Speech: The Diverse Oneness of Writing Systems.* Honolulu: University of Hawaii Press.

[9] Schmandt-Besserat, D. (1992). *Before Writing, Volume 1: From Counting to Cuneiform.* Austin: University of Texas Press.

[10] Ferreiro, E. (1978). What is written in a written sentence: A developmental answer. *Journal of Education*, 160,25-39.

[11] Ferreiro, E. & Teberosky, A. (1982). *Literacy Before Schooling.* London: Heinemann.

[12] Ferreiro, E. (1985). Literacy development: A psychogenetic perspective. In D. Olson, N. Torrence, & A. Hildyard (Eds.), *Literacy, Language and Learning*, pp. 217-228. Cambridge: Cambridge University Press.

[13] Levin, I. & Tolchinsky Landsmann, L. (1989). Becoming literate: Referential and phonetic strategies in early reading and writing.*International Journal of Behavioral Development*, 12, 369-384

[14] Tolchinsky Landsmann, L. & Levin, I. (1985). Writing in preschoolers: An age-related analysis. *Applied Psycholingusitics*, 6, 319-339.

[15] Tolchinsky Landsmann, L. & Levin, I. (1987). Writing in four- to six-year-olds: Representation of semantic and phonetic similarities and differences. *Journal of Child Language*, 14 127-144.

[16] Melkman, R., Tversky, B., and Baratz, D. (1981). Developmental trends in the use of perceptual and conceptual attributes in grouping, clustering and retrieval. *Journal of Experimental Child Psychology*, 31, 470-486.

[17] Clark, E. V. (1973). What's in a word? On the child's acquisition of semantics in his first language. In T. E. Moore (Editor), *Cognitive Development and the Acquisition of Language*.pp. 65-110. NY: Academic Press.

[18] Hughes, M. (1986). *Children and Number: Difficulties in Learning Mathematics*. Oxford: Blackwell.

[19] Neurath, O. (1936). *International Picture Language: The First Rules of Isotype*. London: Kegan Paul, Trench, Trubner & Co., Ltd.

[20] Tversky, B., Kugelmass, S. and Winter, A. (1991) Cross-cultural and developmental trends in graphic productions. *Cognitive Psychology*, 23, 515-557.

[21] Kugelmass, S. & Lieblich, A. (1970). Perceptual exploration in Israeli children. *Child Development*, 41, 1125-1131.

[22] Kugelmass, S. & Lieblich, A. (1979). The impact of learning to read on directionality in perception: A further cross-cultural analysis.*Human Development*, 22, 406-415.

[23] Lieblich, A., Ninio, A., & Kugelmass, S. (1975). Developmental trends in directionality of drawing in Jewish and Arab Israeli children. *Journal of Cross-Cultural Psychology*, 6, 504-510.

[24] Goodnow, J. J., Friedman, S. L., Bernbaum, M., & Lehman, E. B. (1973). Direction and sequence in copying: The effect of learning to write in English and Hebrew. *Journal of Cross-Cultural Psychology*,4, 263-282.

[25] Clark, H. H. (1973). Space, time, semantics, and the child. In T. E. Moore (Ed.), *Cognitive Development and The Acquisition of Language*. pp. 27-63. New York: Academic Press.

[26] Cooper, W. E. & Ross, J. R. (1975). World order. In R. E. Grossman, L. J. San, & T. J. Vance, (Eds.), *Papers From the Parasession on Functionalism*.Chicago: Chicago Linguistic Society.

[27] Lakoff, G. & Johnson, M. (1980). *Metaphors We Live By*. Chicago: University of Chicago Press.

[28] Small, J. P. (1987). Left, right, and center: Direction in Etruscan art. *Opuscula Romana XVI:7*, 125-135.

[29] Stenning, K. and Oberlander, J. (In press). Spatial containment and set membership: A case study of analogy at work. In J. Barnden and K. Holyoak (Editors), *Analogical Connections*. Hillsdale, NJ: Erlbaum

[30] Shin, S.-J. (1991). A situation-theoretic account of valid reasoning with Venn diagrams. In J. Barwise, J. M. Gawron, G. Plotkin, and S. Tutiya (Editors). *Situation Theory and Its Applications*. Volume 2, pp. 581-605. Chicago: University of Chicago Press.

5

Automating Procedures for Generating Chinese Characters

John Loustau[1]

Jong-Ding Wang

5.1 Introduction

The Chinese form of written expression is a complex system whose roots lie in Neolithic pictographs[1]. For at least six thousand years this system has developed alongside Chinese culture. In contrast to Western phonetic alphabets, each Chinese character identifies a single word or particle of thought. Historically, the first characters were simplified versions of more representational images which were too cumbersome for daily use. Later more complex ideas were introduced into the written language by combining or modifying images from existing characters to form new ones. In addition, characters were added without a pictographic reference. Nevertheless, these too followed the convention of one word, one symbol. The result is a system of written expression based on visual images. These images add content to writing, providing added dimension to expression.

The magnitudes of the Chinese system are overwhelming. The total number of characters is more than 100,000. Even the commonly used ones number around 20,000. To counter these numbers, each character is composed from a predetermined group of brush strokes. The total number of distinct strokes required to render all characters is a more tractable sixty or so. A measure of the significance of the strokes is that written style is in large part determined by the style

[1]Department of Mathematics and Statistics, Hunter College of the City University of New York, New York, NY 10021

of the chosen stroke set. In turn, each stroke is determined by a combination of brush path and hand/wrist motion. But aesthetic requirements complicate the simplifying effect of the fixed stroke set, since the shapes of strokes appearing in a character are subtly altered to accommodate the presence of neighboring strokes[2].

The very features, which make the Chinese character system a powerful and pleasing means of expression, make it difficult to automate. With current software systems, the production of commercial digital character sets is a tedious process requiring highly skilled professionals. These characters often lack the lyrical quality common in hand painted brush work. Furthermore, the product is frequently inflexible. For instance, it does not support the transformations necessary for animation applications.

In this paper we report on work aimed at alleviating these problems. Our goal was to develop an automated system which lessens the skill and time required to generate characters while increasing the painterly quality of the finished product and the flexibility of the final data structures. We envision this approach increasing the number of character styles available in word processing and desk top publishing systems. We also see these tools providing an average user the capability of making custom sets of digital characters in his own hand style or in the style of a master calligrapher.

We present our work in the following four sections. In Section 2 we discuss why we feel that this work is important. We consider the aspects of traditional Chinese writing which are not available through*Pinyin* or another Westernized rendering of the language. Our aim is to establish the need to expand the level of automated capabilities present in character generating software systems. The third section outlines the nature of the problem. We present the requirements of the written language. We include the magnitudes of the system, and the rules for character formation. An important aspect of character formation is the brush stroke. We include a description of the requirements for a well formed stroke. In addition, we touch upon some of the aesthetic aspects of Chinese calligraphy.

In the final two sections we discuss software issues. Section 4 concerns the software presently available. As these systems are frequently brush stroke based, we focus on this solution. Next, we list the particular problems that we have chosen and present the direction we have taken. We describe the advantages of a brush path based system over brush stroke based software. We present the capabilities of the software we have developed and we show by example the functions which we have implemented.

5.2 The Case for Chinese Characters

Many Westerners misinterpret the purpose of *Pinyin*. It is not a Latinized form of Mandarin whose purpose is the eventual replacement of the traditional system. Rather, it serves as a standard means of communication between the Chinese and Western cultures. We include this section because of this misunderstanding.

One and a quarter billion people (one quarter of the population of the world) use the Chinese system. It is an integral part of Chinese culture. It has endured and developed along with this culture. It is the one written language in a nation with many spoken languages. To the Chinese it is their heritage and their invention.

Even the difficulties inherent in using the system have content to the Chinese. Mastery requires training, study, and self-discipline. In a society where scholarship is valued more than economic success, these qualities are prized virtues. Well-known calligraphers, those whose mastery of the system well exceeds the norm, are heros. Historical figures, whose calligraphic works are exemplary, are held in such high esteem that their biographies have the quality of myth[3][4].

The use of word symbols with visual context adds a dimension to written expression. One way is through the use of stylized pictographs as single symbols and as elements in more complex characters. In Figure 5.1(a-g) we show a few simple examples. Figure 5.1a is the symbol for 'sun'. The image is derived from a filled circle or disk. The squared boundary is easier to render than a circle. The fill is reduced to a single horizontal stroke. The resulting character is more convenient for daily use and contains the essential elements of the original pictograph. In Figure 5.1b three sun symbols are combined. This character means 'bright'. The character shown in Figure 5.1c means 'star'. In this case the sun symbol sits atop the verb 'give birth'. In turn the give birth symbol contains a modified version of the symbol for 'person'. The character in Figure 5.1b shows the representation of an abstract concept derived from the character for an object. The character for star includes a reference to legendary mythology for the astronomic objects. These last two characters are

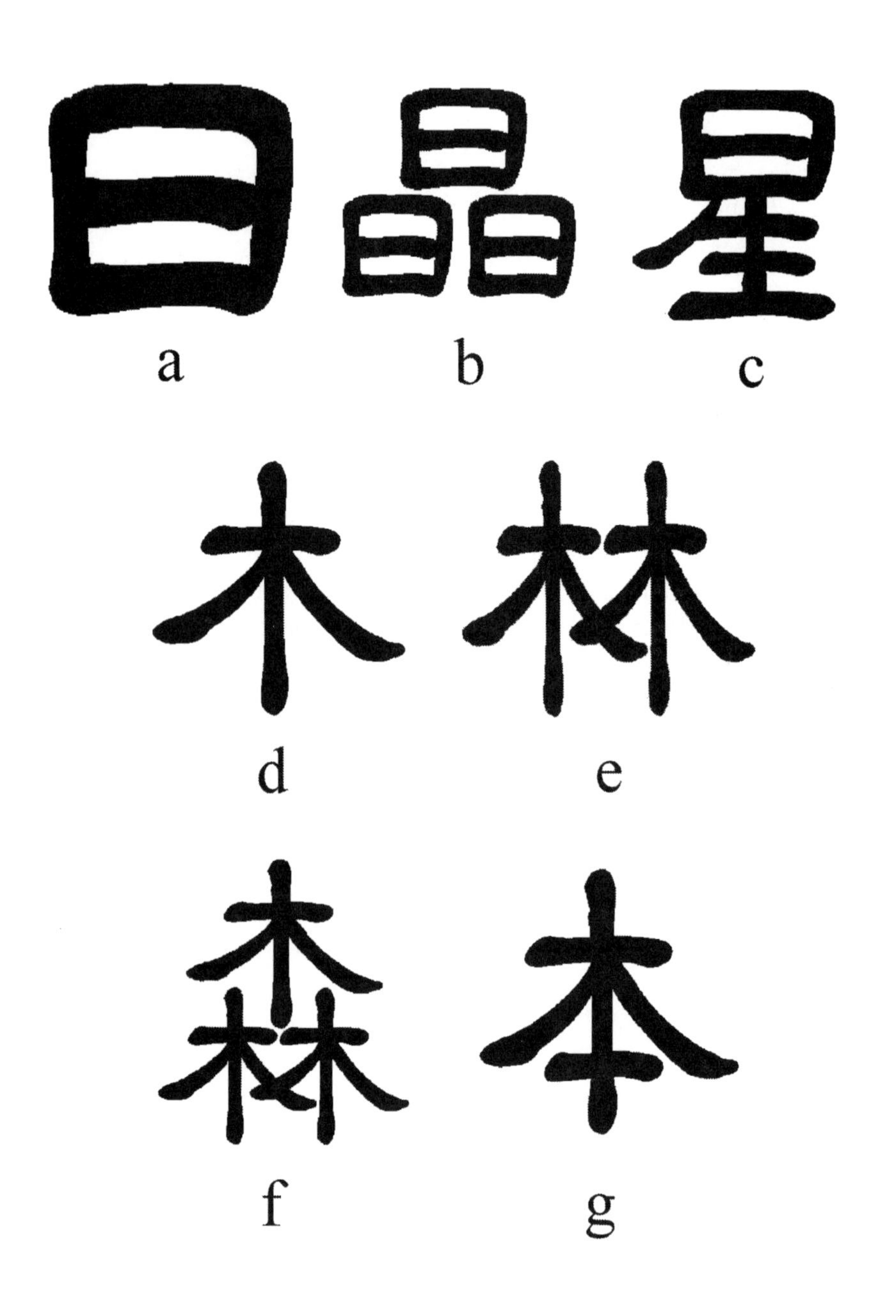

FIGURE 5.1. Characters derived from pictographic sources.

examples of compounds. The compound is only visual. It is not auditory. Both examples show how the character imparts substance to the language through visual images.

Figures 5.1d through 5.1g are derived from a pictograph of a tree. In 5.1d the symbol stands alone to form the character for 'tree'. When the symbol is doubled, it represents 'grove' (Figure 5.1e) and tripled it means 'forest' (Figure 5.1f). Figure 5.1g shows the tree symbol with a horizontal stroke near the base. The device draws attention to the lower part. This modified character represents 'root'. Subsequently, its meaning has been extended to include 'origin'. In this example a symbol for an abstract concept has been derived by modifying the pictographic based symbol for an object.

A second technique for enhancing meaning in written form is through style. Figures 5.2 through 5.4 show varying written styles. In each case there is a correspondence between the look of the work and the content. Figure 5.2 shows a portion of a autobiographical essay by Huai-Su, a Tang Dynasty (A.D. 618-906) calligrapher. The mood is flamboyant. Figure 5.3 is a copy of the *Lan Ting Hsu* by Wang Hsi-Chih (A.D.353), perhaps the most renowned of all calligraphers. The essay extols the feelings after a spring day spent in the company of friends. The feel of the written work is pleasant contemplation. The more formal quality of Figure 5.4 suits a work commissioned by an emperor. In each case the work is further enhanced with subtle and pleasing flourishes. The fourth character in the fifth column (right to left) of Figure 5.3 shows an interesting adaptation. Wang has drawn the character. But the flourishes allow the character to be simultaneously viewed as a shadow. With this perspective the actual character lies in the negative space above each stroke.

We do not mean to imply through these last examples that a given calligrapher routinely alters his style to suit the document he is rendering. A calligrapher, like any artist, chooses subjects which suit his strengths.

5.3 Introduction to Chinese Characters

In this section we consider the technical and aesthetic requirements of the Chinese written language.

We begin with the magnitudes of the system. There are more than 100,000 known characters. This figure includes some examples known

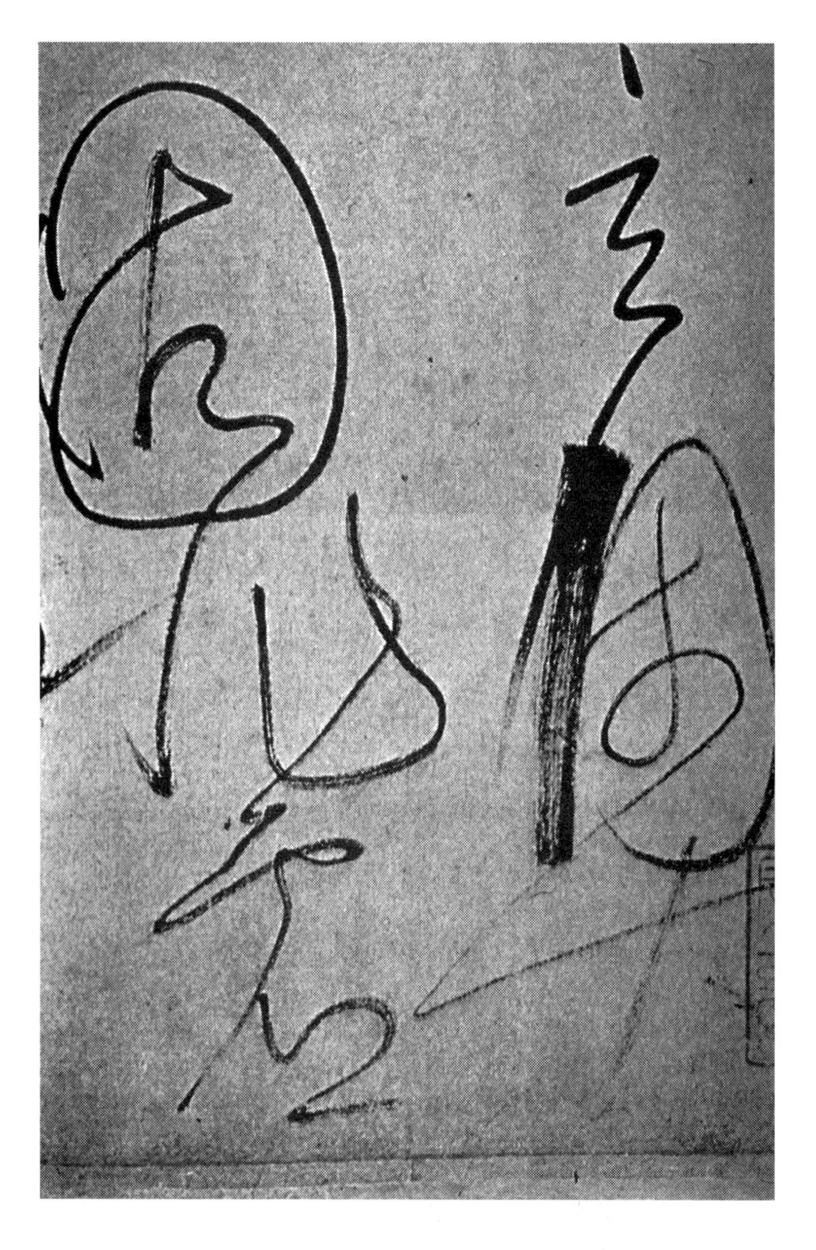

FIGURE 5.2. Huai-Su detail from the *Tzu Hsu Tieb*. (Document is property of the National Palace Museum, Taipei, Taiwan, R.O.C. and is used with kind permission.)

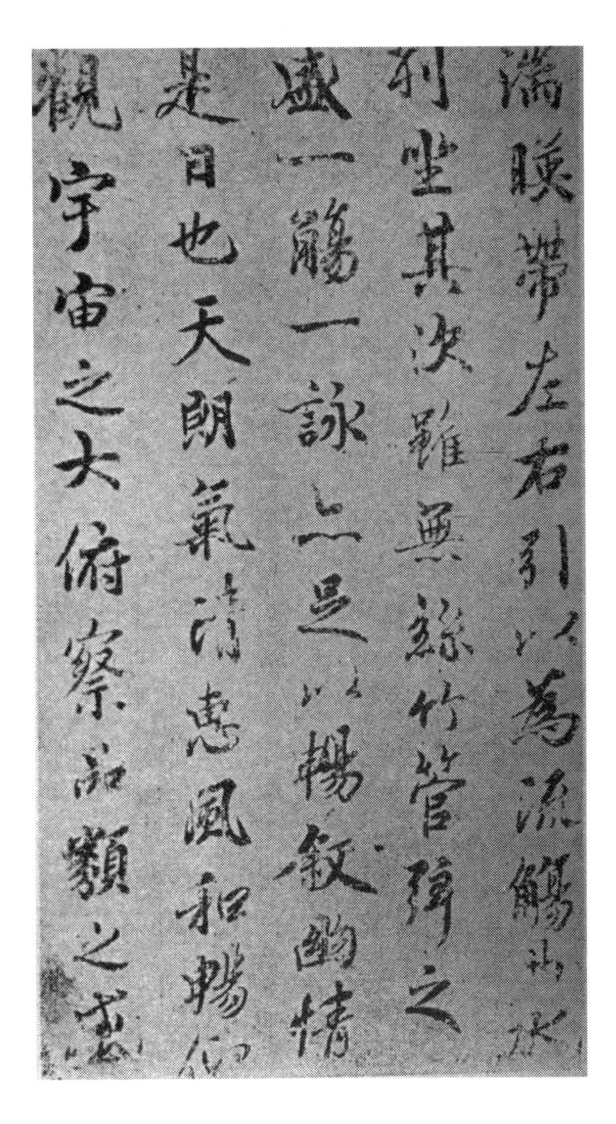

FIGURE 5.3. Wang Hsi-Chih (copy by Chu Shui-Liang): detail from *Lan Ting Hsu*. (Document is property of the National Palace Museum, Taipei, Taiwan, R.O.C. and is used with kind permission.)

only from archaeological study. There are about 40,000 characters in normal usage with about 20,000 of these considered common. Hence, 20,000 characters is minimal for any packaged digital character set. The characters are constructed from brush strokes selected from a stroke set. An individual character may require only one stroke or as many as fifty. Characters with about fifteen strokes are common.

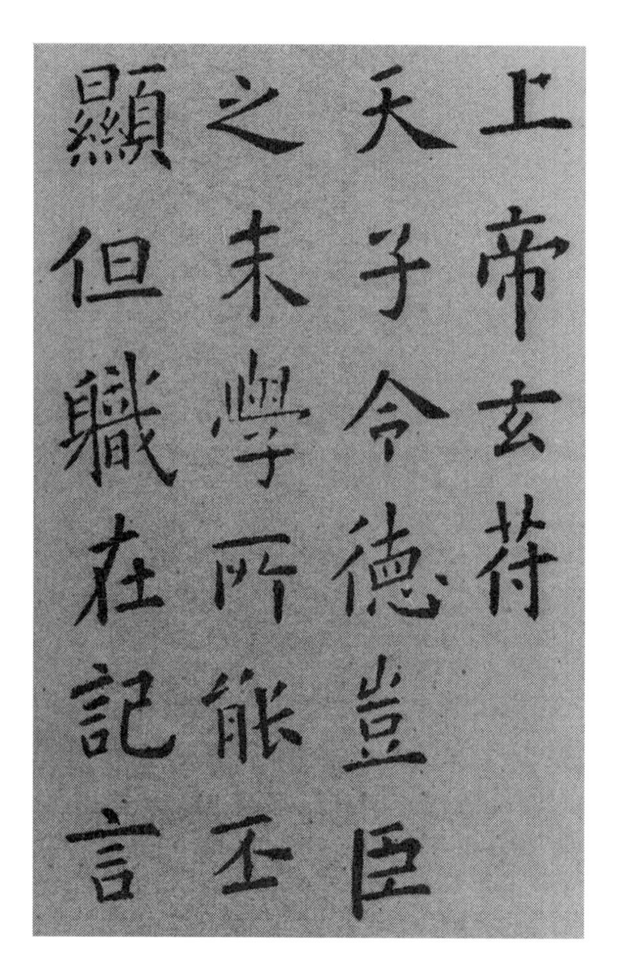

FIGURE 5.4. Hsun Ou-Yang: detail from the *Chiu Cheng Kung*. (Document is property of the National Palace Museum, Taipei, Taiwan, R.O.C. and is used with kind permission.)

There is no agreement on the total number of strokes required for the written language. Some authors list as few as eight while others list two hundred fifty. The difference arises from what is considered a distinct stroke and what is considered a variation. Lists of about sixty strokes are common. A stroke is determined by a designated brush path and hand/wrist instructions such as twist, press and lift. Figure 5.5a shows the character for tree. In the succeeding illustrations the five individual strokes are shown. The inclusion of the brush path highlights the individual strokes one at a time.

The rules of written style address the position of characters and their composition. The characters in a text should occur in straight rows or columns. Each character should fill an imagined rectangle of approximately the same size. Individual characters must display left/right balance. The lower portion must give the impression of a solid supporting base. To achieve balance and support, the strokes used in a character must be modified. Shortening, extending, bending and rotating are common modifications. This requirement makes it impractical to consider a software system that merely defines characters as positioned strokes from a fixed stroke library.

Each character is not only a linguistic unit but also a self-contained visual unit. The strokes used to compose a character form an ordered set. They are positioned and shaped to follow one to the next. The viewer's eye naturally moves about the character passing sequentially through the strokes. His attention does not leave the character before he has noticed each of its components. To accommodate this, strokes must be subtly altered.

Each stroke has three elements: the start, the body, and the end. The start is generally a knob-like appendage created by twisting the brush. The end may also be a knob or it may be pointed. There are several different pointed effects. Pointed effects result from lifting the brush slowly from the paper so that fewer and fewer bristles contact the surface. The stroke body is straight or curved. Its width is uniform. To achieve a particular written style an artist may vary the start and end. Generally the stroke body remains constant. Chinese students learn to make each stroke by learning the brush paths and the hand/wrist instructions. These instructions control the start and end treatments, the stroke bend and thickness.

Chinese calligraphy is usually classified into five basic styles: the *Syao Chyan, Li Shu, Kai Shu, Hsing Shu,* and *Zhao Shu.* Figure 5.6 shows the same text written in four of the five styles. Figures 5.1 and 5.7 are examples of *Zhao Shu.* Each of these styles divides into several substyles. There are works as well which defy easy classification in

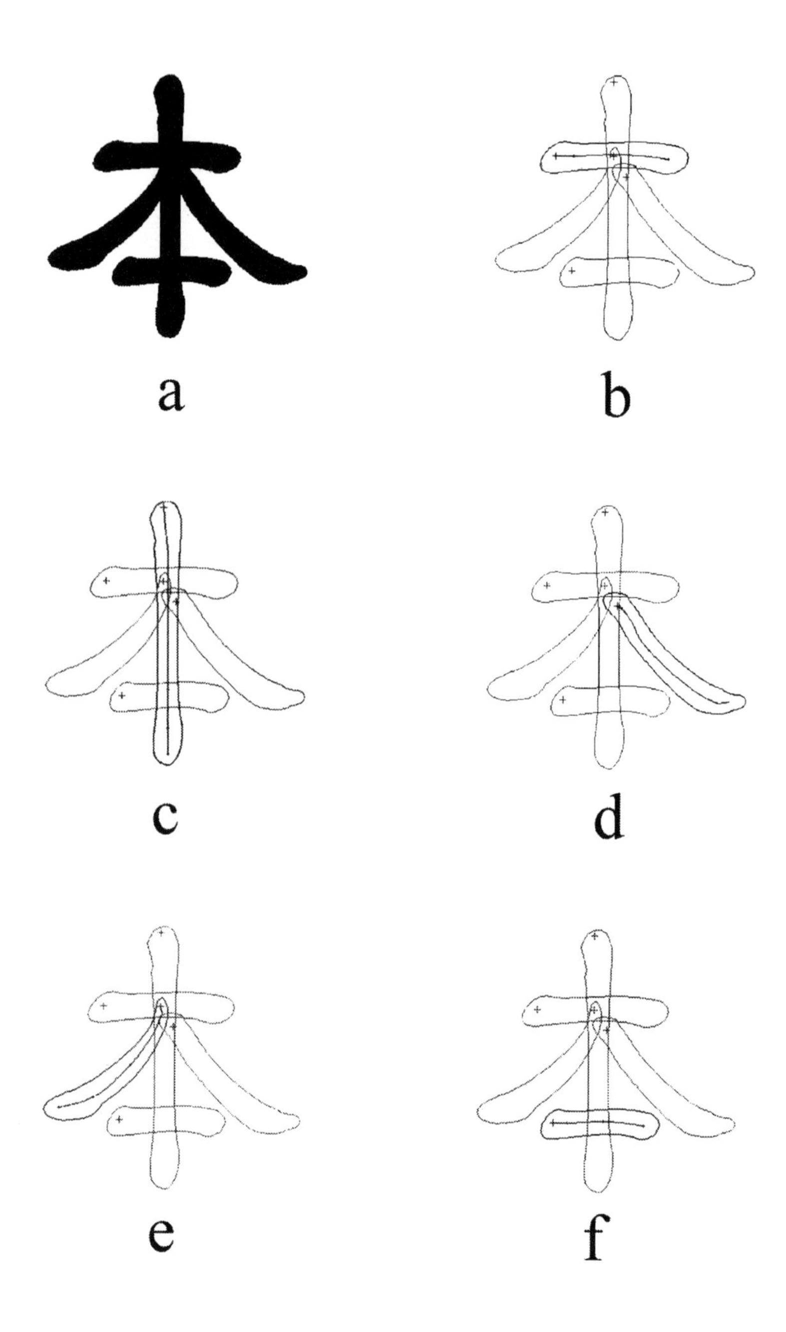

FIGURE 5.5. A character comprised of the five brush strokes.

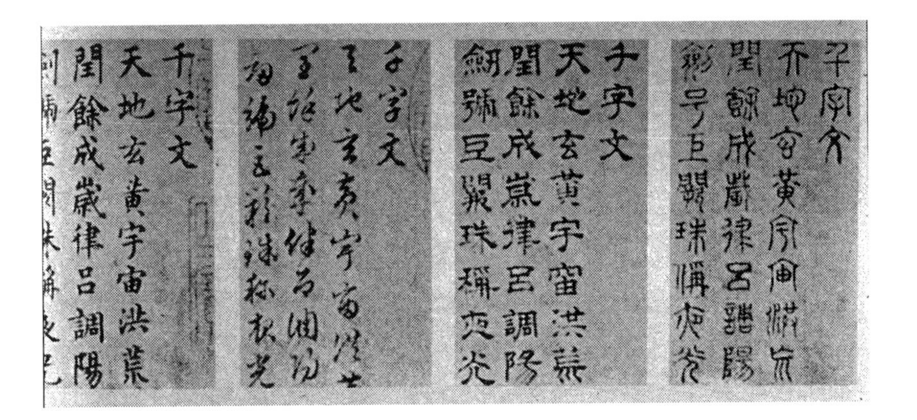

FIGURE 5.6. Wen Cheng-Ming: *Chien Tzu Wen* in *Syao Chyan, Li Shu, Hsing Shu*, and *Kai Shu* (right to left). (Document is property of the National Palace Museum, Taipei, Taiwan, R.O.C. and is used with kind permission.)

this stylistic system.

The *Syao Chyan* is the oldest of the styles. It was developed for carving seals and predates the widespread use of brush and ink on paper. It is most frequently seen in the signature seals which adorn brush calligraphy.

The *Li Shu* is the oldest of the brush styles. It dates from the time of the standardization of the written system. It is recognized by the triangular shapes at the ends of characters. This style is more exacting and formal than the other brush styles. The *Kai Shu* is similar to the *Li Shu*. Each stroke and character is individually rendered. However, now the characters are more loosely developed. Rectangular shapes are not really rectangular. Intersections between strokes are often implied rather than executed. This is the style used in every day correspondence. It is the style taught in school.

In the *Hsing Shu* and the *Zhao Shu* the strokes no longer have distinct individuality. In *Hsing Shu* strokes are often joined one to the next. In the process some strokes are reduced to small wrist movements. In *Zhao Shu* both strokes and characters are frequently connected. Characters are connected one to the next by joining the final stroke of one to the initial stroke of the next. The freedom granted the artist using *Zhao Shu* is so great that the work is often unreadable. Indeed, it is common for works in this style to be accompanied by a rendering in a more legible style. In *Zhao Shu* the visual image has taken precedence over the written content.

There are two aspects of Chinese culture which form the basis for aesthetic judgement of calligraphy. First, all children spend some years learning to write with brush and ink. Therefore, everyone has a basic understanding of the process. They easily recognize special talent. Second, there is great respect for authority. Work that is once accepted as outstanding by the critics does not often fall from favor. These works become a measure to gauge subsequent pieces. In fact, critical appraisal of calligraphy extends back thousands of years. Current discussions of calligraphy often quote critics who are ancient by Western standards.

There are specific terms used when discussing calligraphy. We list a few. The work should be *vibrant* and *vital.* This effect occurs in spite of the fact pieces are executed in a careful and studied manner. The work may be *polished* or *unpolished.* These terms distinguish between smooth edged or rough edged brushing. The work in Figure 5.1 is unpolished while the one in Figure 5.7 exemplifies polish. The work should be *clever.* In particular the artist's technique should contain novel and interesting effects. We have already noted the rendering of shadows in Figure 5.3. The work should have an *academic* quality. The artist should demonstrate his understanding of the historical context. The work should be *elegant* as recognized in Chinese terms. It should be *heavy.* This term is explained as being too great for the paper upon which it is painted. It refers to the overwhelming feeling we have when experiencing a great work like Mozart's "Requiem" or Monet's "Water Lilies". Finally, it should be serious. This term has special meaning to the culture.

5.4 Current Software Capabilities

In this section we review some current software solutions. These systems have been developed as tools used by firms which sell character sets, as commercial products for ordinary users and as implementations of academic studies.

Systems used by companies which generate character sets are in large part slightly modified 2-D drafting systems. The main thrust of the system is to provide the user with unlimited image modeling capabilities. However, little attention is paid to the peculiarities of the context which would aid in the process.

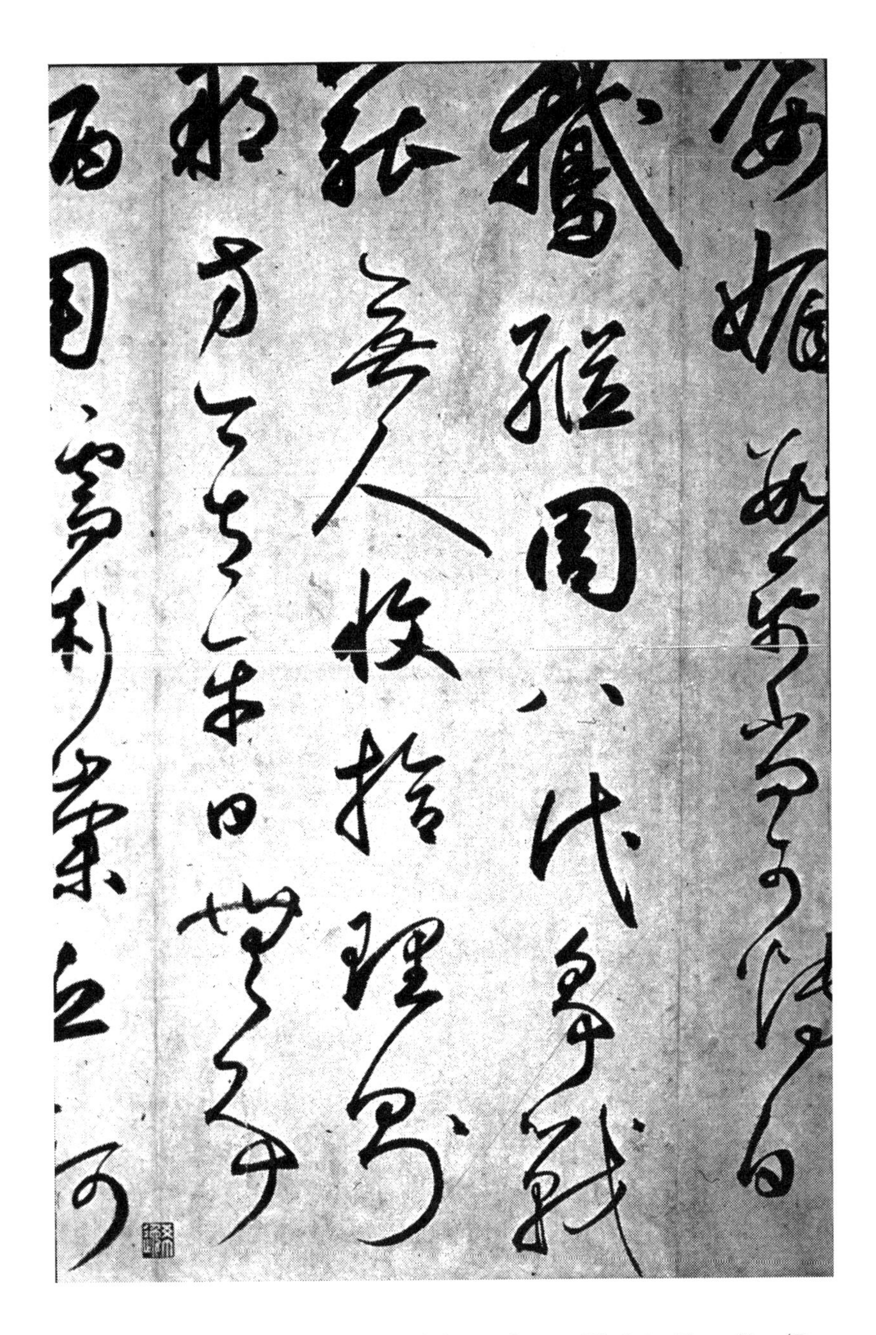

FIGURE 5.7. Hsueh Shao-Peng: Detail from a letter. Work in *Zhao Shu*. (Document is property of the National Palace Museum, Taipei, Taiwan, R.O.C. and is used with kind permission.)

These systems are stroke based. Strokes are prestored as a sequence of connected Bezier curves. These curves are developed using 2-D Bezier drafting software with standard functionality including mouse or pen input, scaling, knot point, and tangent point editing. Characters are subsequently constructed from the stroke set by a recall and drag procedure. The strokes are then modified by reappling the same 2-D drafting functions. This produces an image similar to the large central image in Figure 5.11.

To finish the character, the strokes are filled. The filled image is then stored in bit map format. If a character outline is required, the screen image of the filled character is scanned to locate the character boundary. This image is likewise stored in bit map fashion. The output is adequate for word processing and desk top publishing applications, the primary market for character sets.

A second level of software is intended for the nonprofessional user. These applications are available as off-the-shelf products. They vary greatly in their capabilities. Some are purely bit map based. The operator constructs new characters by combining elements from an initial set of characters provided with the software. Image editing is restricted to altering screen bits and scaling. Other examples have the feel of systems used to manipulate Western font sets. They are Bezier based but lack the full functionality required to support written symbols as complex as those used by the Chinese. There is software that does produce brush strokes from brush paths and permit subsequent modification of the stroke through path based commands. Used in conjunction with other software modules, this product provides capabilities similar to the system described below. However, the mode of access is limited at important stages in the process. Different modules are not fully integrated. Work done in one module and then modified in another is no longer accessible in the prior one.

An entirely different approach is proposed by Strassman [5]. His work is brush-path based and uses B-splines. He describes a mouse as the preferred input device. However, with current technology a pressure sensitive pen would be preferred. The pen path would define the brush path and the pressure would determine the stroke width. The brush stroke results from B-spline curves developed along perpendiculars to the brush path. The B-splines represent hairs on a brush. The area between the curves is filled using anti-aliasing techniques. The output is a finished brush stroke which looks remarkably like ink on absorbent paper applied with a round brush.

In a subsequent work, Pham [6] achieves results similar to Strassman's while using a simpler model. Again the brush stroke is developed from the brush path as B-splines defined along perpendiculars to the path. However, in this approach the brush hairs are three dimensional curves where the first two coordinates denote position and the third ink flow.

To our knowledge neither work has been realized in a commercial product.

5.5 The Path Based Bezier Curve Approach

In our work we have focused on two issues, the time and expertise required to generate high quality characters and the adequacy of the character data structures. In the first direction we have developed a more powerful means to interact with a character generating system. In the second, we have achieved a geometric representation of the character. This geometric representation permits a full range of subsequent processing.

The illustrations in this section as well as Figures 5.1 and 5.5 are output from our software. In accordance with our goal, these characters are in a classic style. In particular, most of the strokes are modeled after the the work of Chu I-Tsun. To further emphasize our goal, these characters are generated by the authors, neither of whom is experienced at this work.

Our approach uses Bezier curves to model brush strokes used in Chinese calligraphy. We chose Bezier curves because stroke boundaries have isolated points where the curve is not smooth[7]. In particular, the stroke outline is a continuous curve which may be locally not differentiable or differentiable but not continuously differentiable. For instance, the pointed-end effect of a practised calligrapher produces a singularity. In addition, when the artist changes direction, he may halt his hand motion for an instant. The result can be a curve which is only singly differentiable. Amongst the curves commonly used in computer graphics, Bezier curves are the simplest curves that have the properties just described.

The interactive stroke generating module is based on the brush path. The path is built first. Then the stroke is initiated from the brush path. Modeling is done using hand/wrist instructions. This approach is convenient from the operator view point as these instructions are the very devices learned in school. Therefore, the system is based on widely held knowledge. Access is not restricted to trained professionals. It is equally important that the brush path provides a basis for a powerful set of instructions. This approach, which appears conceptually meaningful, also makes good engineering sense.

The brush path is created using standard 2-D modeling functions. First, data points along the path are input with a mouse or pen. Then using well-known procedures, we initialize the path as connected Bezier segments. Refinement is achieved by modifying the locations of the Bezier guide points. Since the brush path is a simple arc, this step is significantly simpler than modeling the corresponding stroke boundary.

The stroke module focuses on the brush path. The initial stroke outline is programatically generated from the path. At this time a software procedure logically ties the outline curve to the path curve. Through the association of points on the path to points on the stroke outline, the software can locate the beginning, end, top, bottom and the left, right of the stroke boundary as well as segments along the stroke body. The operator refines the stroke shape using commands such as light brush, heavy brush, twist end, point end, stretch, compress, bend, rotate, and scale. These commands are executed at path points and echoed to the stroke. Most commands have one sided and two sided variations to achieve asymmetric as well as symmetric effects.

The advantage of path based operations is that the operator can easily identify regions of the stroke or locations along the stroke to the software. The path is a 'handle' which allows the operator to affect both sides of the stroke at the same time. The result is to double the effectiveness of modeling operations.

In addition we provide the operator with direct access to the stroke boundary. These commands permit modification of knot point and tangent point locations. This access is usually required only for fine tuning. In many cases it is unnecessary.

In Figures 5.8a through 5.8d we illustrate stages during the modeling of a brush stroke. Figure 5.8a shows three curves. The inner one is the brush path. The center curve shows a digitized outline of a hand drawn brush stroke. This is the target. The outer image is the initial brush stroke derived from the path. In its initial form this curve has uniform distance from the path. This distance is determined either as a function of the brush path length or the operator input. In Figure 5.8b light-brush commands have been applied along the length of the path. The effect is to draw the curve in toward the target. Next point-end and twist-end operations are applied at the ends. The result is shown in Figure 5.8c. Figure 5.8d shows the final stage. The asymmetrical effect of the twist at the beginning of the stroke is achieved with a one sided command. To achieve perfect overlap, 2-D modeling operations are applied to both ends.

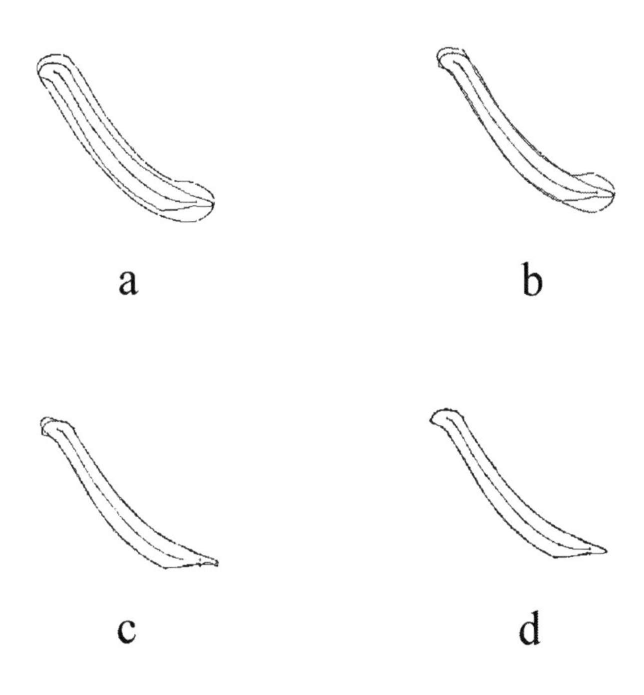

FIGURE 5.8. Stroke Modeling shown in stages.

The character is constructed by recall and drag. For final modifications, the operator applies the same path-based operations just described. Examples of strokes that were modified at this stage occur both in Figure 5.1 and in Figure 5.11. For instance, the third stroke (top to bottom) on the left hand side of the character of Figure 5.11 has been shortened by shortening the center portion of the path. This operation is distinct from scaling. Using this function, the stroke can be shortened without changing the size of the twists.

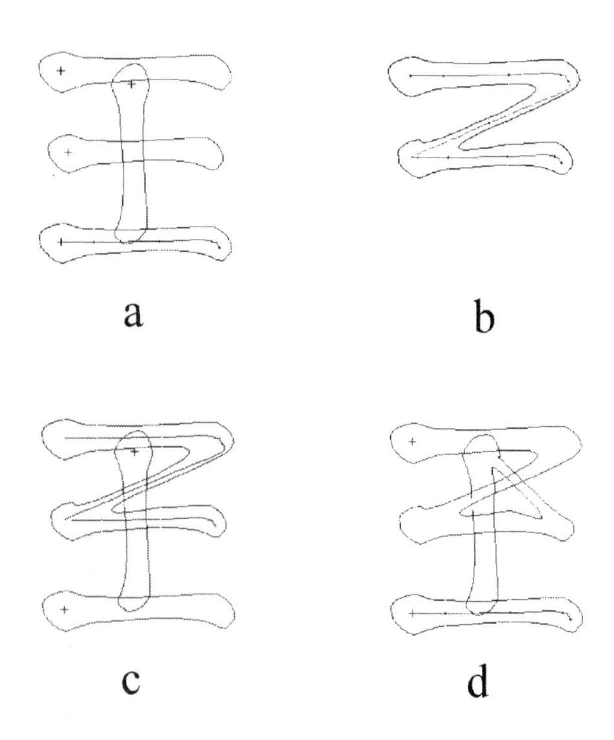

FIGURE 5.9. Transitions from regular style to *Li Shu*.

To generate *Hsing Shu* or *Zhao Shu* from one of the other brush styles, a system must have a function which joins separate strokes. In the former case strokes within a character are joined. In the latter case, strokes from separate characters are connected. Figure 5.9 shows the steps in the transition of a character in regular style to *Hsing Shu*. With the aid of the brush path the transformation is a straightforward process. The operator chooses the strokes. The software connects the corresponding brush paths. Using the new path as a guide, the two strokes are opened and joined as new bounding arcs for the resulting single stroke are formed. The operator can make finishing touches to the result. This was not done in this example.

Our second area of interest is the data structures for a character. In particular, the data structure for characters should be sufficiently flexible to support a full range of subsequent processing. Since the desired functionality is predominantly geometric, then the data structure should be a geometric one. Indeed, a bit map representation is devoid of any geometric information. Such a representation, though adequate to display images, cannot support transformations such as scaling and rotating. Thus, the goal of a flexible character representation leads to the goal of a geometric representation.

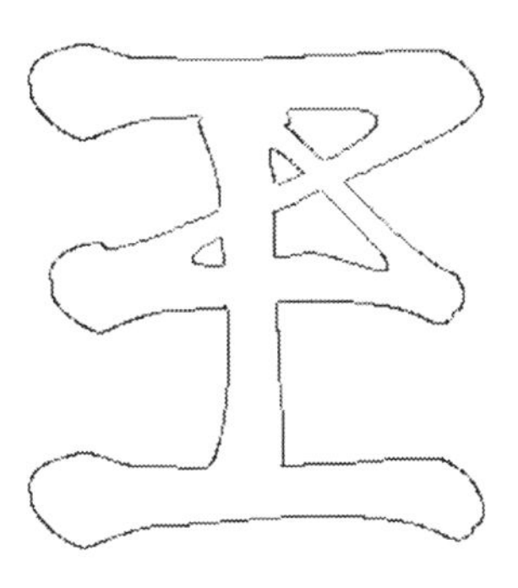

FIGURE 5.10. The *Hsing Shu* character shown in outline form.

A geometric representation for the final character requires that we resolve the bounding curves of the character to yield a single Bezier representation. This in turn requires that we locate the intersections of the stroke outline curves and determine which of the intersecting stroke curves determines the character outline.

The stroke intersection problem is complicated by a fact about written style which we did not find mentioned in the literature. Strokes tend to meet where they are straight. In geometric terms intersections occur at points where the stroke boundary curves have low curvature. The result is that standard techniques to compute Bezier curve intersections do not yield reliable results. We solved the intersection problem using techniques borrowed from ray-tracing. Although our procedure would not be adequate for an engineering application, it is sufficiently precise for the problem at hand. Most importantly, it provides a basis for a fully reliable automated process.

For the second problem, the determination of the character outline curve, we turned to techniques from hidden surface removal. The brush path is useful at this stage as it gives ready access to interior points for each stroke.

Figures 5.10, 5.11, and 5.12 show characters rendered in outline

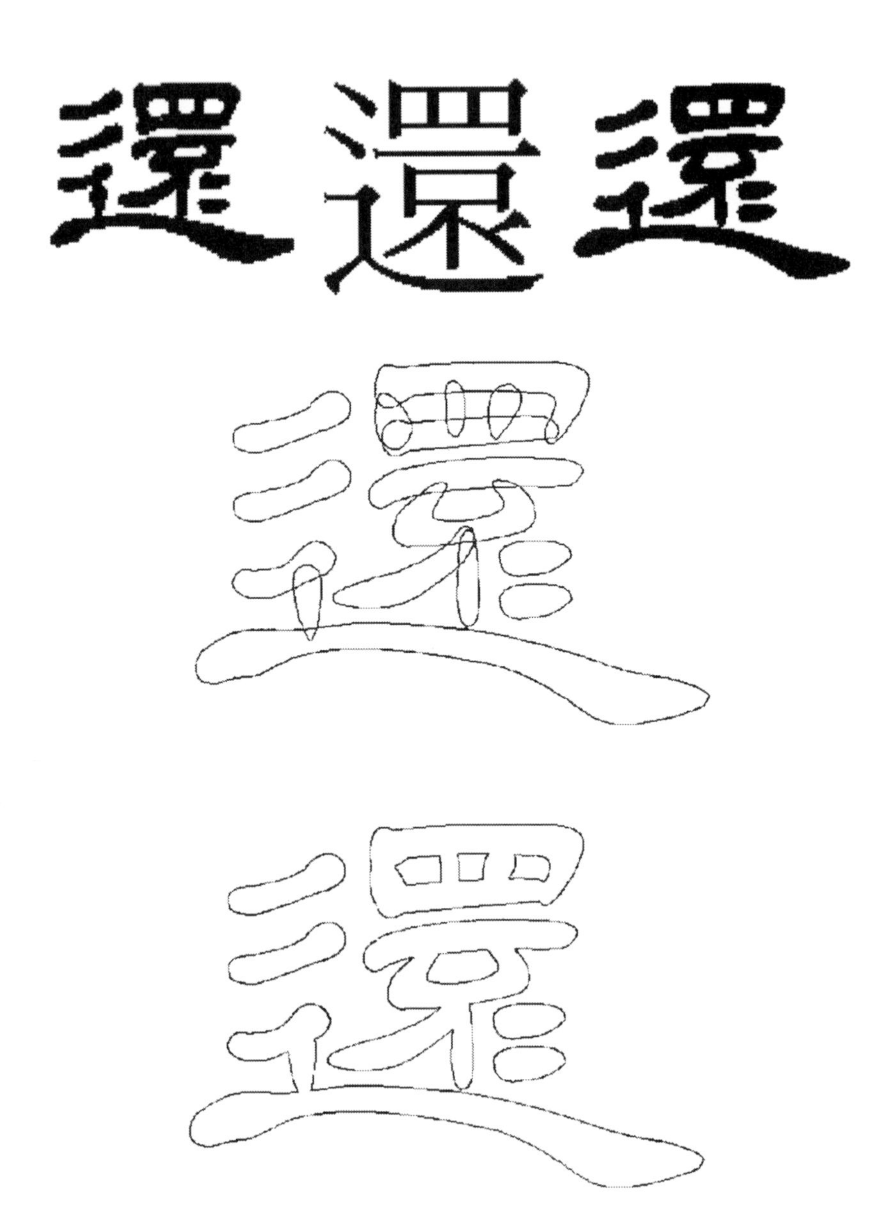

FIGURE 5.11. A character as joined strokes and as character outline.

form. The small character in the upper-left-hand corner of Figure 5.11 is a digitized image from a work by Chu I-Tsun. The middle character of the top row is the corresponding character from a commercially available character set. The large central character in this figure is the same character generated by our software. This image illustrates the stage after recalling, positioning, and modifying strokes. The large bottom image shows the character in outline form. Looking back at the version with stroke outlines, notice the central area where the curves from three strokes intersect at nearly the same point. There is an additional complication here as two of the strokes are tangent at their intersection. Complex situations such as this are common in Chinese written style. The character at the upper left is derived from the bottom image using a shrink and fill. The geometric representation makes this operation routine. Note the remarkable similarity between the computer generated image on the upper right and the original at the upper left. Altogether, the character outline procedure functions in interactive time.

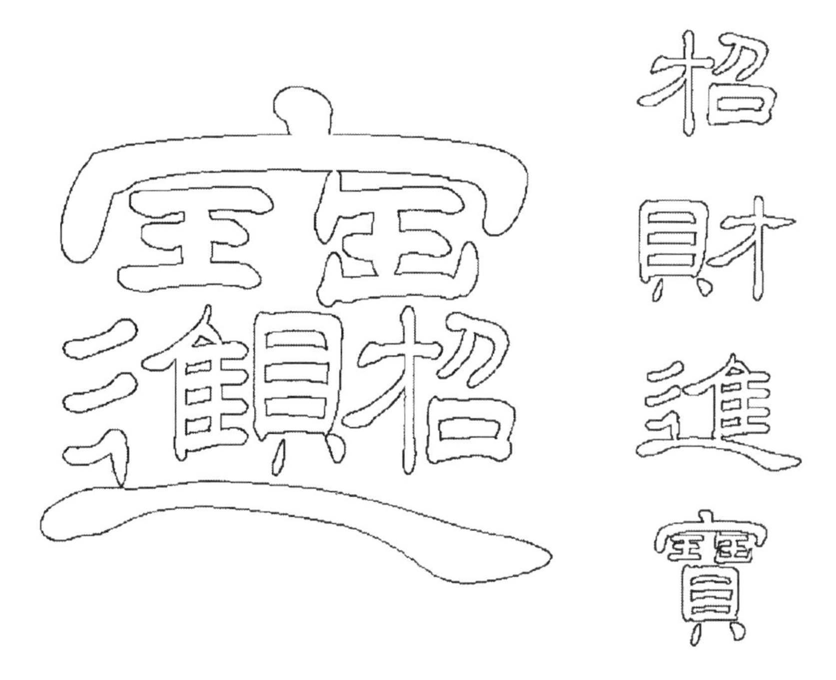

FIGURE 5.12. A new years greeting as individual character and as a combined character.

5.6 References

[1] Kwo, Da-Wei (1981). *Chinese Brushwork: Its History, Aesthetic and Technique.* Montclair, NJ: Allenheld, Osmun and Co.

[2] Yee, Chiang (1973). *Chinese Calligraphy: An Introduction to its Aesthetic and Technique.*Cambridge, MA: Harvard University Press.

[3] Chen, Chih-Mai (1966). *Chinese Calligraphers and their Art.* Melbourne, Australia: Melbourne University Press.

[4] National Palace Museum (1992). *Catalogue of the Special Exhibition of the Beauty of Calligraphy.* Taipei, Taiwan, ROC.

[5] Strassman, S. (1986). Hairy Brushes. *SIGGRAPH Proceedings 1986.*New York, NY: ACM, pp. 225-232.

[6] Pham, Binh (1991). Expressive Brush Strokes. *CVGIP: Graphical Models and Image Processing*, 53,1-6.

[7] Loustau, J., and Dillon,M. (1993). *Linear Geometry with Computer Graphics.* New York, NY: Marcel Dekker Inc.

6

Multimedia Representational Aids in Urban Planning Support Systems

Michael J. Shiffer[1]

6.1 Introduction

Many urban planning situations involve informal queries about the conditions of a physical environment by groups of people. The queries are typically posed as questions such as "What is there?" or "What would it be like if . . .?". These queries may be supplemented with a variety of information in various combinations such as maps, narrative descriptions, photographic images, and human gestures (such as hand waving or pointing). An increasing number of these queries are directed at computer-based planning support systems. Planning support systems (PSS) are computer applications that allow urban and regional planners to access a variety of tools for analytic purposes (such as forecasting population shifts, traffic patterns, etc.). Unfortunately, many of today's PSS lack the descriptive abilities of images and human gestures. A set of tools known as representational aids now make it possible to link descriptive images, such as digital video and sound, to information that would be otherwise represented quantitatively.

This paper will discuss the relevance of representational aids to PSS, particularly how they describe the physical environment. This will be accomplished through a discussion of the theoretical underpinnings of representational aids and illustrated with examples drawn from work in progress. Some of the examples used are taken from a prototype PSS designed to help the National Capital Planning Commission (NCPC) assess the environmental impact of major

---

[1] Planning Support Systems Group, Department of Urban Studies and Planning, MIT, 77 Massachusetts Ave., Room 9-514, Cambridge, MA 02139.

developments on the nation's capital. A more in-depth example will be provided through the description of representational aids incorporated in a prototype PSS designed to support the exploration of re-use alternatives for Chanute Air Force Base in Rantoul, Illinois.

6.2 Challenges to Urban Planning Support Systems

The ability to communicate is central to the process of urban planning. A Planning Support System (PSS) may be of little use if the relevant items pertaining to a problem cannot be communicated to, and among, a group of people. Many of today's PSSs lack the descriptive abilities of images and human gestures. For example, while a spreadsheet model may predict changes in population density, it is not able to provide an example of how crowded the streets will be. Similarly, quantitative measurements used to describe levels of automobile traffic or aircraft noise may be meaningless to the lay person who possesses the responsibilities of making decisions based on this information.

This lack of communication can also lead to shifts in the power structure of many organizations. For example, the difficulties encountered in mastering the application of these tools often cause less technically-oriented individuals to be excluded from the planning process as control over planning support systems "tends to increase the power of administrators, technical experts, and technically sophisticated groups at the expense of those who lack the expertise to use them effectively" ([1], p. 4). This can lead to less than satisfactory outcomes to planning problems as well as political, ethical, and psychological difficulties for both the planners and the planned-for[2][3].

There has been some movement towards the objective of making PSSs more accessible. For example, it has been noted that the development and rapidly expanding capabilities of the microcomputer represent the democratization of computing power in society [4]. In fact, many vertical microcomputer applications have been written for planning support as a result of this democratization. Unfortunately, they contain varying degrees of usability.

Thus, regardless of computing platform, a need exists to make analytic tools, and their outputs, more manipulable, understandable, and appealing so that information that would normally be meaningless and intimidating to the lay person can be comprehended. This would in turn carry the "democratization" of computing several steps further so that rather than having a proliferation of tools in the hands of individuals, we can have a proliferation of tools that individuals are able to use effectively. Several technological trends are making it possible to move in this direction. This paper will focus on those that fall under the category of "representational aids."

6.3 Addressing the Challenges: Representational Aids

An improvement in the understandability of PSS, (and consequent improvements in their impacts on the planning process), is made possible through the employment of representational aids. Representational aids have historical roots that go far beyond the development of computer-based tools. Images such as simple physical maps, photographs, and even cave paintings have all acted as representational aids. These have historically possessed the ability to clarify what would otherwise be difficult to convey using other means of communication such as language or gestures. In the context of this paper, we will focus on the representational aid as a means of clarifying human computer interaction in the urban planning environment.

Computer-based representational aids are designed to allow the human to interact with the decision-aiding algorithms in as natural a way as possible [5]. This minimizes the cognitive load that can be imposed by an analytic tool, since it reduces the need for the user to translate his/her ideas into those understandable by the computer and vice-versa.

Through the use of examples, representational aids illustrate proposed changes to an environment based on information drawn from comparable situations. For example, a representational aid can illustrate what a Boeing 747 aircraft will sound like taking off from airport X if there is a representation of a 747 takeoff at airport Y and if airport Y is sufficiently similar to airport X. One could extrapolate comparable examples with other environmental phenomena such as levels of traffic congestion, population density, etc.

Representational aids come in two forms: those that make it easier for the human to send commands to the computer, and those that make the computer's output more understandable for the user.

Norman [6] characterizes these two tasks as bridging the Gulfs of Execution and Evaluation. The Gulf of Execution refers to the gap between the user's intentions and goals, and the inputs recognized by the computer. For example, you may want the computer to present you with a list of the land uses for a community, but unless you know the syntax and commands necessary to convey this request to the computer, you're out of luck. We will refer to the tools that help translate your intentions to the computer, thus bridging the Gulf of Execution, as "Input Aids". The Gulf of Evaluation refers to the gap between the computers output and the users perception of the problem domain. For example, a table of numbers printed from a computer that represents map coordinates and land-use codes for the aforementioned community may be difficult to understand. We will refer to the tools that can translate the coordinates and codes into a graphic land use map, thus bridging the gulf of evaluation, as "Output Aids".

6.4 Input Aids

A recent technological trend that addresses the need for more effective human-computer interaction has been the movement from verbal or "command-driven" to visual or "gestural" human-computer interfaces. Many previous computer interfaces were command driven. That is, the user had to translate his or her intentions into commands that could be understood by the machine. This usually entailed a large amount of memorization that could be equated with learning an entirely new "language." Unfortunately, those who could not communicate in this language were either completely left out of computer-aided processes or heavily reliant upon those who spoke the computer language.

Gestural interfaces have been developed to overcome the need to memorize commands by translating the user's actions into commands that can be understood by the machine. This is accomplished by providing a graphical interface displayed in a form that matches the way one thinks about a problem. This way the user can make "rapid incremental reversible operations whose impact on the object of interest is immediately visible" ([7], p. 91). For example, if I wished to recast a phrase during the writing of this paper (using an on-screen pointing device such as a mouse), I could select the phrase in question by performing the gesture of dragging a representation of a highlighting pen over the phrase.

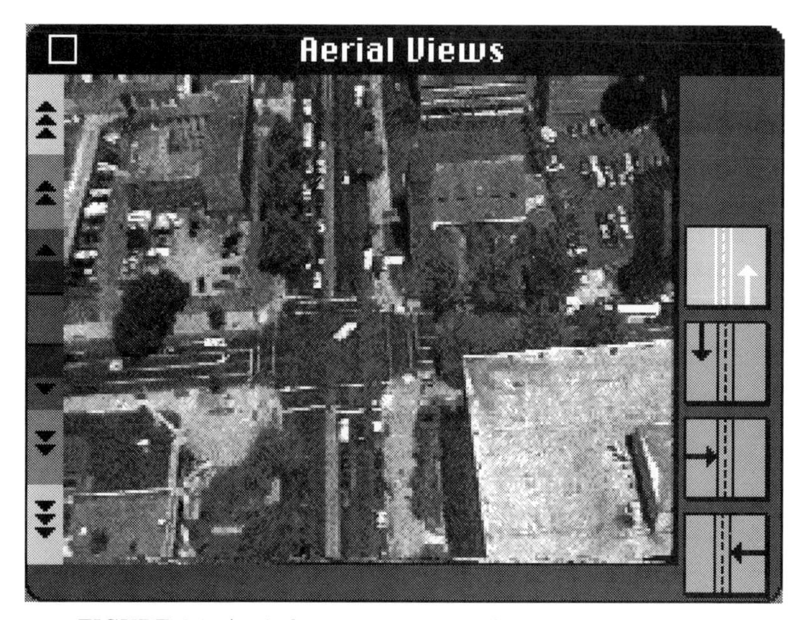

FIGURE 6.1. A window containing a video navigation tool.

Another example of a gestural interface can be provided in the context of a navigation tool that makes use of digitally stored aerial video images. In Figure 6.1, the video image represents an oblique view of a selected street from an altitude of 500 feet. The controller at the left edge of the figure allows the user to control the direction of flight (forward or reverse), as well as the speed of flight, by sliding the pointer towards either end of the continuum.

The user can determine the camera angle by selecting one of the iconic buttons at the right side of the figure. The arrows on the icons represent the direction the camera was pointing with respect to the subject, (in this case, the subject is the street).Furthermore, the subject itself, that is the route flown, can be determined by selecting it from a map. Thus, in this context, the user has the ability to use a gestural interface to interact with algorithms designed to display any one of thousands of digital video images randomly accessible through the computer. The result is the illusion of flight along a particular path. Thus, rather than implementing analytic models by typing cryptic code or numbers, planners can point to maps and photos, slide graphs and bars, and push buttons using a gestural interface in order to elicit a response from the computer.

6.5 Output Aids

Representational aids can also influence the output characteristics of the machine so that they may be more readily understood by the users. This involves designing the displays so as to present information in a manner that makes the human-machine interaction so engaging that the machine essentially becomes "transparent" to the human. That is, the human interacts directly with the information in deference to the vehicle that is providing the information. So, ideally, humans become so engaged in the activity of working out their problems that they forget that they are using a computer. Such engagement involves incorporating concepts such as "visual presence" and "multiple representations" [6]. Visual presence aids in evaluation by providing visual reminders of what was done. For example, grayed-out text could be used to signify portions of a plan that were already read. Furthermore, visual presence can aid in interpretation by providing such things as words, graphs, moving images, and pictures.

An example of visual presence is provided with the multiattribute evaluation tool illustrated in Figure 6.2. The tool is designed to enable an individual or group to evaluate several urban development alternatives using multiple criteria. It uses a form of multiattribute analysis based on a Simple Multiattribute Rating Technique, (SMART) described by Edwards [8]. The interface for the evaluation mechanism uses digitized video and sound combined with direct manipulation graphics. The digitized video is used to identify each alternative. Scores and weights are input by sliding a graphic bar. Sound can also be added so as to better describe an alternative's performance on a given attribute. Feedback, or visual presence, is provided by an arrow that continually moves from one alternative to another indicating a recommended choice as user preferences are changed.

Multiple representations of a problem [9] enable the user to view information in several different contexts thus offering the potential to generate alternative approaches to a problem.

Figure 6.3 illustrates the use of multiple representations to illus-

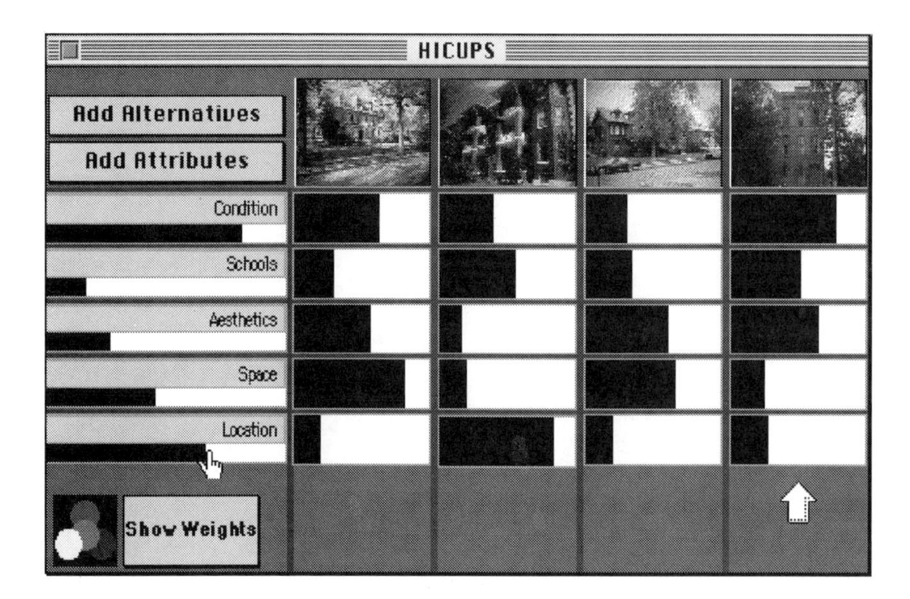

FIGURE 6.2. An example of visual presence where an arrow moves from one alternative to another indicating a recommended choice as the user's preferences are changed by sliding graphic bars with a pointing device.

trate average daily traffic counts (ADT) for automobiles on selected streets in Washington, D.C. ADT is represented in the traditional numeric form; it is also represented graphically, with a bar; dynamically, with a clip of digital motion video; and audibly with the level of traffic noise played back at the level experienced in the field. While the bar graph may look redundant in the static image portrayed in Figure 6.3, its utility becomes apparent as users point to different streets on the projected map, causing the bar to fluctuate. In this manner, users can compare relative levels of traffic to one another more easily.

Through the use of images and the implementation of concepts such as gestural interfaces, visual presence, and multiple representations of data, planners and decision makers can better understand the physical environment relevant to an issue at hand. To gain a better understanding of how these concepts can work together it may be useful to take a closer look at the application of representational aids in the context of a more in-depth example.

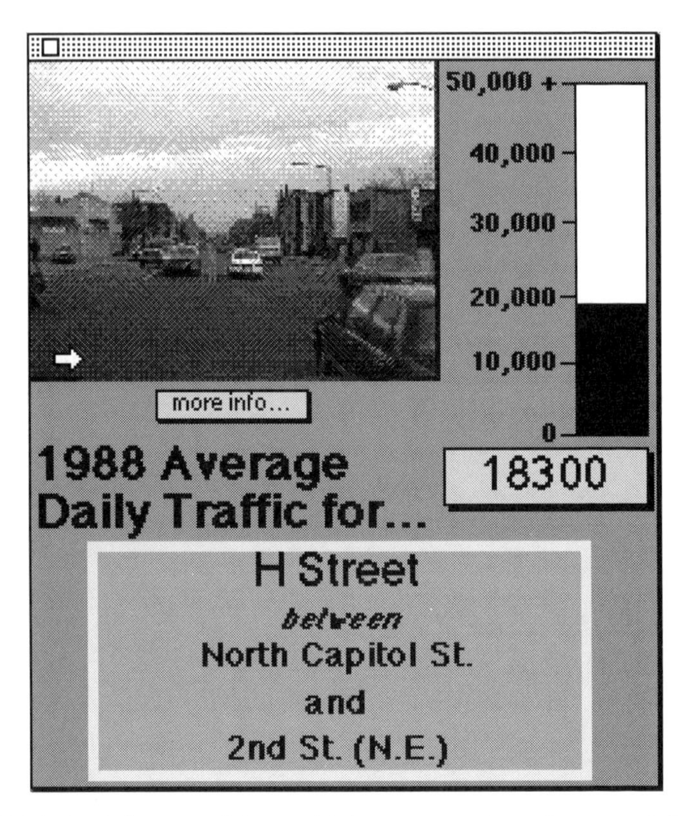

FIGURE 6.3. A window containing multiple representations of average daily traffic data.

6.6 Application of Representational Aids: Rantoul, Illinois

An example of the extensive use of representational aids was illustrated in Rantoul, Illinois where village officials experimented with a prototype PSS in the context of exploring re-use alternatives for Chanute Air Force Base. The system's users included the mayor, community development director, several village board members, and other interested parties. The system was demonstrated in a group environment where the users focused their attention on a single projected display screen.

Chanute, which was scheduled to close, had not experienced aircraft flight operations in nearly twenty years. Yet one of the potential re-use alternatives was a repair facility for a major commercial air-

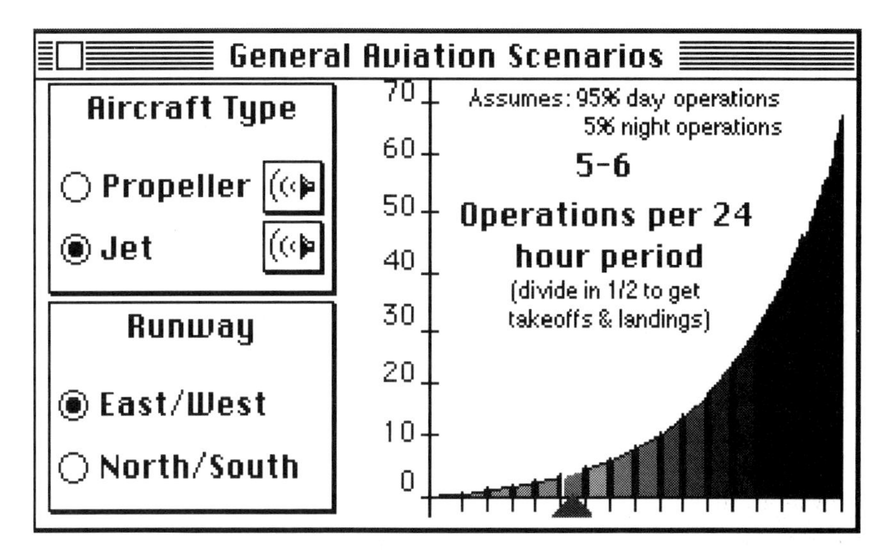

FIGURE 6.4. The input interface for the general aviation component of the noise impact visualization tool in the Rantoul PSS.

line. In order to describe some of the impacts of aircraft operations, representational aids were employed in several contexts, one of which was the employment of a tool to illustrate the discrete sound component of an aircraft noise impact measure so that the user could select a location on a map and experience what an aircraft takeoff would sound like if the user were standing outdoors at the selected location.

This was created by recording takeoffs and measuring decibel levels at various geographic points in the vicinity of an airport that provided similar conditions to those at Rantoul/Chanute. The noises were then digitized and linked to loudness information stored in the computer. Before the system's use, the noises would be calibrated with a decibel meter to match actual observed conditions. They could then be accessed in the context of a larger PSS designed to support collaborative planning activities [10][11].

The discrete noise occurrences that were illustrated above were part of a more comprehensive measure of noise impact known as LDN. Simply stated, LDN is an average of noise frequency and intensity over a time [12]. Assessments of LDN impacts were accomplished by allowing the users to set the frequency of operations, direction of operations, runway selection, and type of aircraft by directly ma-

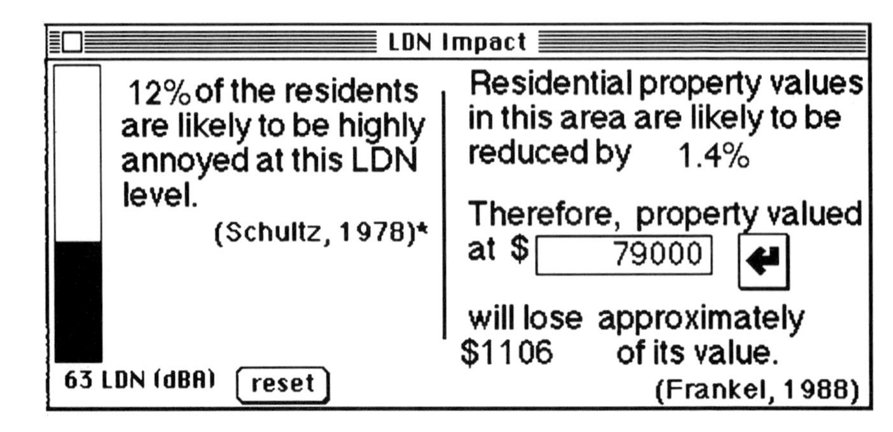

FIGURE 6.5. A window that illustrates the impact of an aircraft noise scenario in Rantoul, Illinois.

nipulating the control panel illustrated in Figure 6.4. The control panel translates the users' desires to the computer that, in turn, displays appropriate LDN contours on a base map as transparent color polygons.

The representation window, illustrated in Figure 6.5, incorporates representations of LDN's impact upon property values [13] as well as the approximate number of residents that are likely to be highly annoyed by LDNs at various levels [14]. Since LDN is a continuous measure, an interpolation tool has been built into the representation window that allows the user to study approximate impacts of LDNs between contours. The interpolation is accomplished by sliding a bar that represents LDN values.

These interpolations are immediately reflected in the visualization measures (property value and annoyance impacts). The noise impact visualization tool also allows the users to calculate the assessed values for areas falling within LDN contours either automatically for aggregate measures, or manually for specific properties.

At several points during the implementation of the system, the benefits of combining a collaborative situation with representational aids were evident. For example, a lively conversation about Chanute's former flight patterns took place after visualizing a scenario of jet operations on the North-South runway. From this conversation came speculation that if the north-south runway were used, planes would need to turn rapidly after take-off to expedite noise abatement in the

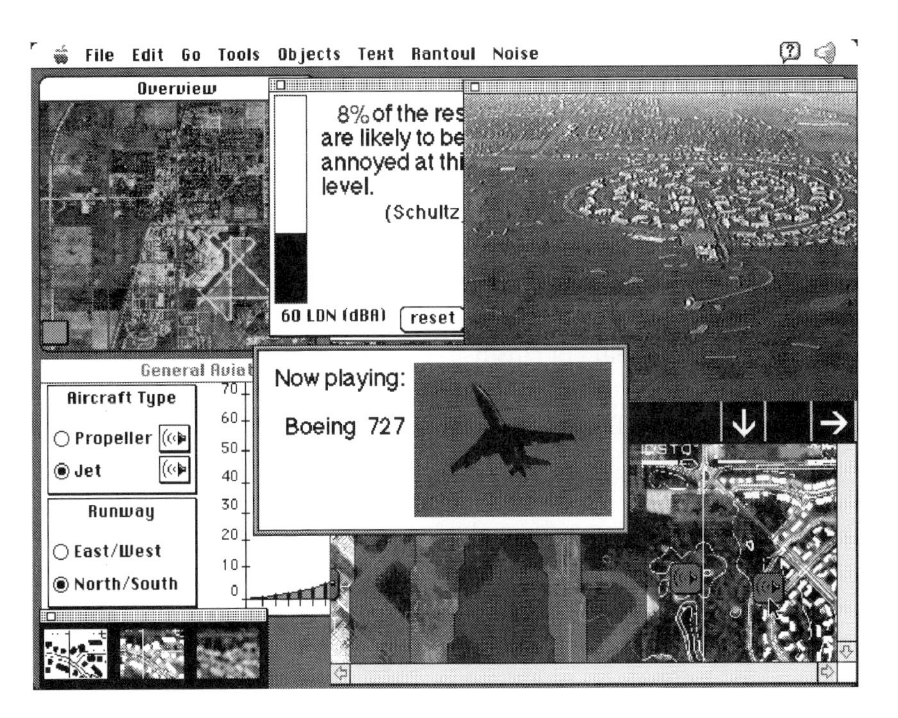

FIGURE 6.6. Multiple representations provided by the noise imapct visualization tool.

community. Participants then gave examples of other airports where that type of action had been taken and the system provided an illustration of what the flight would look like from the pilot's point of view using oblique aerial video images. In this manner the participants were able to (1) hear the takeoff, (2) see the affected areas on a map, and (3) see the affected areas from the point of view of the pilot. This was all provided through the system's ability to display multiple representations (see Figure 6.6).

Thus, the users were able to visualize or "audioize" the relative impacts of different scenarios with the help of a multimedia interface thereby highlighting the benefits of representational aids for analytic tools. This widened the range of scenarios that could be explored significantly since the users were not restricted to the outputs of one or two computer runs or restricted to the use of a difficult-to-understand tool.

6.7 Conclusions

Through the use of representational aids, planning support systems (PSS) can empower groups and individuals who have traditionally been informationally disadvantaged due to a lack of technical sophistication, thereby allowing more people to become involved in the generation and evaluation of alternatives. Exactly who benefits from such empowerment will depend on the situations in which the PSS is implemented.

The relative benefits of this use of representational aids in decision support systems have been supported by psychological evidence. For example, Brill, et al. [15] noted that an effective interface was apparently so powerful that the interface itself encouraged greater problem exploration (p. 755). This supports the claim that "the graphic knowledge system concept for decision support is based most simply on the empirical finding that visualization enhances human problem solving" [16].

Representational aids will not completely replace quantitative measures of environmental phenomena. Rather they will serve to supplement such measures through multiple representations. Where accuracy and precision are important, the "free use" of a representational aid may be inappropriate as it may serve to cloud otherwise objective information with subjective images.

A popular criticism of simple to use interfaces has been the adage "ease of use promotes abuse". While it is possible that easier access to tools and information can offer easier access to their misuse, this research has illustrated that ease of use can also promote experimentation and exploration. Such experimentation and exploration may result in the identification of related issues to a particular problem, or the generation of new alternatives.

Finally a caveat is in order. Just as these tools have the capacity to enable compelling representations of an environment, they have the capacity to illustrate compelling misrepresentations. Whether this is manifested through the naiveté of the user or through less scrupulous intentions, one needs to view these tools with a watchful eye.

6.8 Acknowledgements

Partial support for this project was provided by the Department of Urban and Regional Planning and Apple Seedlings Grants at the University of Illinois. Additional support was provided through National Capital Planning Commission contract # 91-02. Thanks are due to the Village of Rantoul, the National Capital Planning Commission, Lew Hopkins, Joe Ferreira, and Lyna Wiggins.

6.9 References

[1] Klosterman, R. E. (1987). Guidelines for future computer-aided planning models. In *Proceedings of the Annual Conference of the Urban and Regional Information Systems Association IV.*

[2] Rouse, W.B. & Morris, N. M. (1976). Understanding and enhancing user acceptance of computer technology. *IEEE Transactions on Systems, Man, and Cybernetics*, 6, 965-973.

[3] Forester, J. (1980). Critical theory and planning practice. *Journal of the American Planning Association*, 46, 275-286.

[4] Brail, R.K. (1987). *Microcomputers in Urban Planning.* New Brunswick, N.J.: Center for Urban Policy Research.

[5] Zachary, W. (1986). A cognitively based functional taxonomy of decision support techniques. *Human-Computer Interaction*, 2, 25-63.

[6] Norman, D. A. (1986). Cognitive engineering. In Norman, D.A. and Draper, S.W. (Eds.), *User Centered System Design: New Perspectives on Human Computer Interaction.* Hillsdale, NJ: Lawrence Erlbaum, pp. 31-61.

[7] Hutchins, E. L., Hollan, J. D. & Norman, D. A. (1986). Direct manipulation interfaces. In Norman, D.A. and Draper, S.W.(Eds.), *User Centered System Design: New Perspectives on Human Computer Interaction.* Hillsdale, NJ: Lawrence Erlbaum, pp. 87-124.

[8] Edwards, W. (1977) How to use multiattribute utility for social decisionmaking. *IEEE Transactions on Systems, Man, and Cybernetics*, 7, 326-340.

[9] Rasmussen, J. (1986). *Information Processing and Human Machine Interaction: An Approach to Cognitive Engineering.* New York: North Holland.

[10] Lang, L. (1992). GIS comes to life. *Computer Graphics World,* 15 (10), 27-36.

[11] Shiffer, M.J. (1992).Towards a collaborative planning system. *Environment and Planning B,* 19,709-722.

[12] Ford, R.D. (1987). Physical assessment of transportation noise. In Nelson, P.M. (Ed.), *Transportation Noise Reference Book.* London: Butterworth, pp. 3-25.

[13] Frankel, M., (1988).The impact of aircraft noise on residential property markets, *Illinois Business Review* 45(5), 8-13.

[14] Schultz, T.J. (1978). Synthesis of social surveys on noise annoyance. *Journal of the Acoustical Society of America,* 64(2), 377-405.

[15] Brill, E. D., Flach, J. M., Hopkins, L. D., and Ranjithan, S. (1990). MGA: A decision support system for complex, incompletely defined problems. *IEEE Transactions on Systems, Man, and Cybernetics,* 20, 745-757.

[16] Woods, D.D. (1986). Paradigms for intelligent decision support. In Hollnagel,E., Mancini,G. and Woods,D.D.(Eds.).*Intelligent Decision Support in Process Environments.* Heidelberg: Springer-Verlag, pp. 153-173.

7

Gesture Translation: Using Conventional Musical Instruments in Unconventional Ways

Robert S. Williams[1]

7.1 Introduction

One way to define the process of understanding images is as the process of moving information from one mind to another. This paper will address that process as it applies to the problem of moving information from the mind of the designer of a digital musical device to the mind of a musician desiring to use the device in a musically interesting way. In other words, this is a paper on interface design, and will present an interface intended to be appropriate for a musician using digital musical instruments.

The goal for such an interface is to allow a musician as much control as possible over a digital music system. Figure 7.1 shows the components of one such system. This system consists of the musician, a sequencer, and a sound generating device. The sequencer contains prerecorded digital information, and sends this information to the sound generating device, which interprets the information as a sequence of musical events and makes those events audible. The musician can similarly send musical events to the sound generating device by, e.g., playing notes on an instrument that can convert those notes to digital information. But allowing the musician only the ability to send musical events of this kind limits the amount of control possible. A musician might also want control over timbre, dynamics, orchestration, tempo, and other kinds of sound-shaping information. Information of this type can also be sent to a sound

[1]Computer Science Department, Pace University, New York, NY 10038-1502

generating device, and is interpreted as a sequence of control events, which modify how the musical events are made audible.

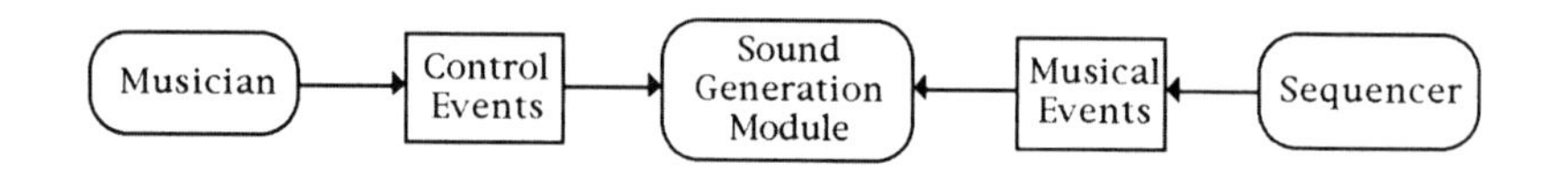

FIGURE 7.1. A digital music system.

Today's digital musician has a seemingly unlimited potential for shaping the music that he or she creates. Digital synthesizers come equipped with hundreds of sounds at the musician's disposal; each of these sounds might have dozens of parameters that can be modified during the course of a performance. The addition of computer based sequencing software compounds the number of choices available: the musician can control not only the dynamics, tempo, orchestration, but even compositional elements of the music being sequenced.

This potential comes at a price. Performers can only do so many things at once; the mere process of generating sounds limits the possibilities for controlling those sounds. Additionally, the potential for controlling sounds does not necessarily translate to the ability to control sounds: if the means of realizing choices are not readily available during the course of a performance, those choices do not exist for any practical purpose.

Ideally, the means to harness the potential of digital instruments should relate to a performer's existing knowledge of their instrument. Performers normally devote years to perfecting a set of skills relating to generating sounds on an instrument. These skills allow a performer to visualize a musical effect and almost instantaneously give that visualization life as a set of physical gestures. If the gestural vocabulary developed in the process of learning an instrument could be adapted to new ends, a performer could quickly realize the potential inherent in digital instruments.

7.2 The Need for Something Better

To illustrate both the possibilities and the limitations of current digital instruments, consider Roland's Sound Canvas sound generation module. The Sound Canvas has sixteen different parts; each part can be assigned one of 317 different sounds. Each sound can be further modified by a variety of parameters, including vibrato rate and depth, cutoff frequency, resonance, and envelope attack, decay, and release time. The level, pitch, and duration of each tone can be further shaped with parameters that include volume, transposition, portamento, pitch bend, and sustain. The module also contains reverb and chorus effects, which can be assigned individually to each of the sixteen parts. The Sound Canvas manual [1] lists a total of sixtynine different parameters that can be modified in various ways.

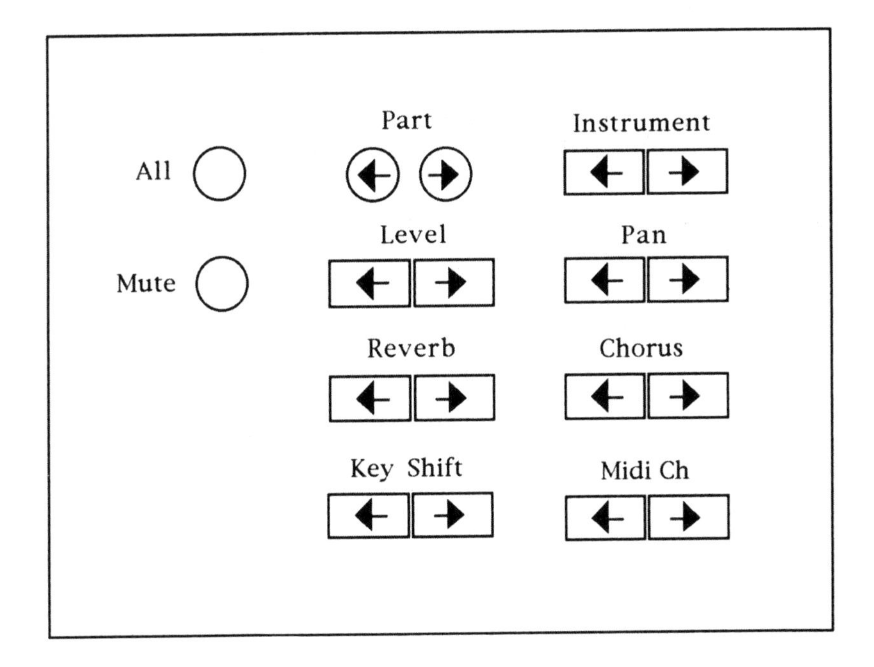

FIGURE 7.2. The front panel of the Sound Canvas.

Figure 7.2 shows an abstraction of the front panel of the Sound Canvas. In addition to the power switch and a knob for adjusting the overall output level of the unit (not shown in the figure), there are a total of eighteen buttons available to a user of the unit. The buttons are mostly arranged in pairs that allow the user to make a selection from a list of items: the left button moves down the list, and the right button moves up the list. For example, to change the sound for part 3, the part buttons would be used to step through the list of parts; then, the instrument buttons would be used to select a new sound.

All parameters can be adjusted from the front panel of the Sound Canvas, but adjustment can be cumbersome. For example, there are 317 sounds available, but only 128 can be selected using the instrument buttons as described above. Each of the 128 sounds immediately available has one or more variations. Variations are selected by first choosing a sound (using the instrument buttons) and then pressing both instrument buttons simultaneously. A single button is then used to select from among the variations.

Some parameters are even more difficult to adjust via the front panel. Figure 7.3 shows the procedure (taken from [1]) for adjusting parameters such as vibrato rate and depth. To change the value of one parameter for one part requires a six-step process, with two steps involving pressing two buttons simultaneously. It should be pointed out that the buttons on the Sound Canvas are very small, and it can be quite difficult to press two of them simultaneously.

We can use the front panel of the Sound Canvas as a starting point for examining desirable means for controlling digital musical instruments in a performance situation. There are essentially three features that are desirable in a controller:

1. Flexibility. Access to a large number of parameters should be possible, and each parameter should be adjustable in isolation. The front panel of the Sound Canvas is extremely flexible, since all parameters are individually adjustable.

2. Ease of use. The number of steps necessary to change a parameter should be kept to a minimum. This is clearly not the case with the front panel of the Sound Canvas. As many as six steps are involved in choosing certain parameters; two of these steps involve pressing two buttons simultaneously, while three steps involve scrolling through lists of items.

3. Compatibility. If the performer is already trained in some aspect of performing, gestures for controlling parameter should

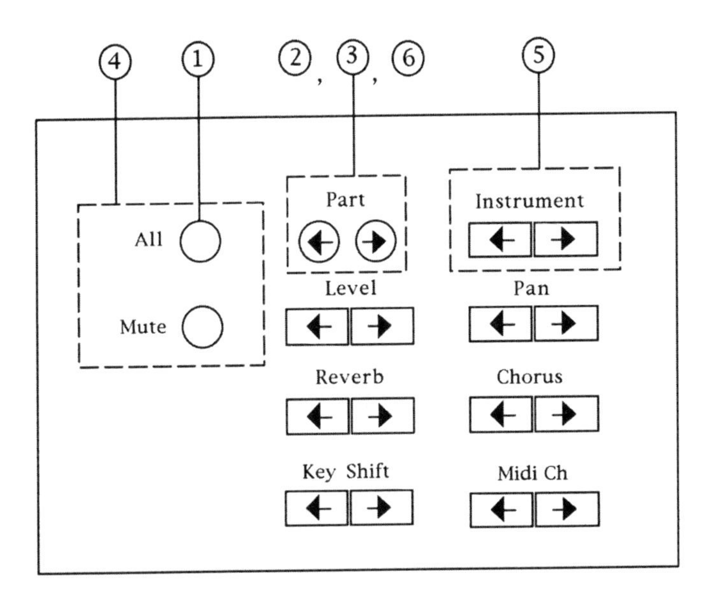

1. Make sure that the ALL button indicator is off. If the button is on press the button to turn it off.
2. Press the PART buttons simultaneously.
3. Use PART buttons to select the part for setting.
4. Use ALL and MUTE buttons to select the sound parameter (vibrato rate, depth; attack time, decay time, release time, etc.).
5. Use the INSTRUMENT buttons to select the value.
6. After setting, press the PART buttons simultaneously to finalize the setings.

FIGURE 7.3. Adjusting parameters using the Sound Canvas front panel.

be compatible with those familiar to the performer. Controlling the front panel of the Sound Canvas involves learning a set of gestures far different from those involved in playing most conventional instruments.

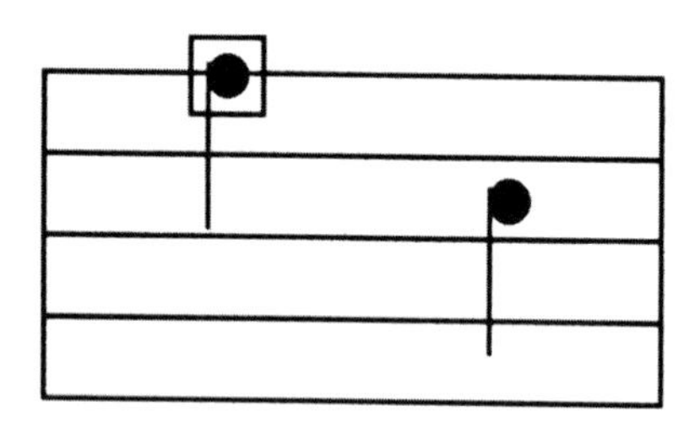

FIGURE 7.4. Adjusting parameters using gesture translation.

7.3 Gesture Translation

This paper will describe an approach to controlling digital instruments, called gesture translation, that allows a musician to understand the control problem in terms of an already familiar domain, namely, that of performance of music on a traditional instrument. With gesture translation, the same gestures used for generating sounds are also used to shape sounds, but not at the same time: one alternates between using a set of gestures in the traditional way (to generate sounds), and using the same set of gestures in non-traditional ways (to shape the sounds).

A musician using gesture translation could use a slightly embellished version of traditional music notation to control sounds as well as to generate them. For example, Figure 7.4 shows one possible way of using gesture translation to describe how to change a parameter on the Sound Canvas. This description calls for the musician to play two notes, and is given in an embellished form of music notation (the box around the first note means that the musician must step on a foot pedal while playing the note). With the proper translation of notes, the process described in Figure 7.4 can be the same process as that described in Figure 7.3. Clearly, the second description is easier to use than the first (it contains fewer steps), and it is also more compatible with a musician's knowledge, since it is given in traditional notation and the process it describes can be carried out using a conventional musical instrument. In principle, gesture translation can also provide as much flexibility as, e.g., the front panel of the Sound Canvas.

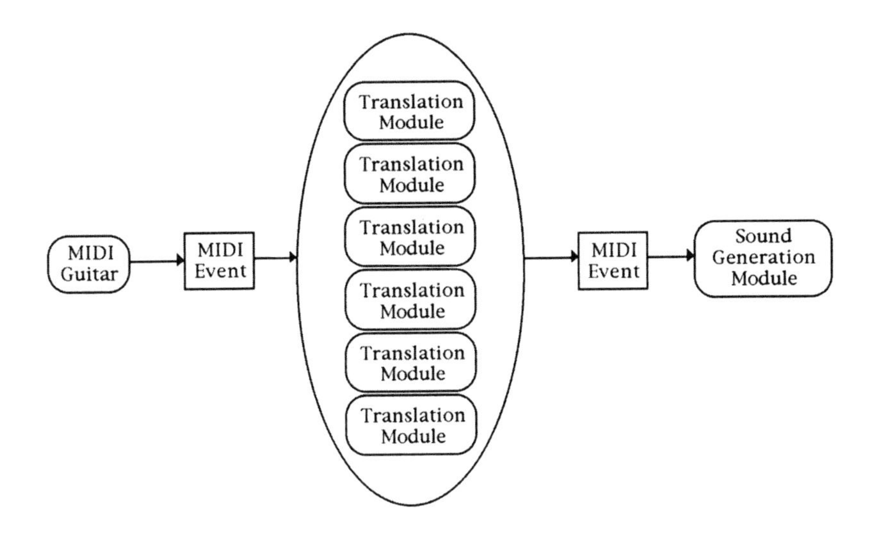

FIGURE 7.5. The gesture translation process using MIDI guitar

7.4 The Current System

The current work in this direction involves developing software that can translate gestures to any of a number of sound generating or sound shaping actions. The software is designed to work with a MIDI guitar (MIDI, or Musical Instrument Digital Interface, is a protocol for sending digital messages representing musical events), and allows each string of the guitar to be treated as a separate sound generating or sound shaping module.

Figure 7.5 illustrates this realization of the gesture translation process. The MIDI guitar allows each note played to be translated to a MIDI event made up of the pitch of the note played, its loudness, and the string on which the note was played. This information is sent as input to six translation modules (one per string on the guitar). If the string of the input event corresponds to a module, the module produces as output a new MIDI event. This is sent to a sound generation module (such as the Sound Canvas) where it produces an action appropriate to the translation module. The simplest action is to play the note given as input: the MIDI event is merely sent verbatim to the sound generation module. Most translation modules process the input in various ways before sending out a new MIDI event. For instance, a translation module for changing sounds might translate its input event into a program change event, with the pitch being translated to a program number (a component of a program change event). When the sound generation module received this new event, it would respond by selecting a sound corresponding to the program number.

The gesture translation process as described so far is somewhat limited in terms of flexibility: only six translation modules are available, so only six distinct types of control are possible. This limitation is overcome by the addition of a translator selection module, a special type of translation module. Instead of producing a MIDI event, a translator selection module uses the pitch of its input event to select a translation module to be associated with the string of the input. In this way, the number of translation modules available to each string is the same as the number of pitches that can be played on the string (between twenty and twenty-four on most guitars). If even more flexibility is desired, additional translator selection modules can be added.

Visual information is displayed to a performer via graphic abstractions of a guitar's fretboard, as illustrated in Figure 7.6. Each such abstraction is a grid containing six rows (representing the strings of a guitar) and twentyfour columns (representing the possible notes that can be played on each string). At the bottom of the display are representations of fret markers, as found on most guitars. Notes not available to a performer are displayed in grey. Notes currently being played are displayed in black. Figure 7.6 represents a fretboard for which twenty-two notes are available for each string. The note at the seventh position on the fourth string (A below middle C in standard tuning) is currently being played.

The system displays three fretboard representations at all times:

1. A display that shows the current choices of the translator se-

lection module. The notes displayed represent the translation modules currently assigned to each string. This is referred to as the function display.

2. A display that shows the output channels that will be affected by each string's translation modules. This is called the channel display.

3. A display that shows which notes are currently being played on each string. This is called the active display.

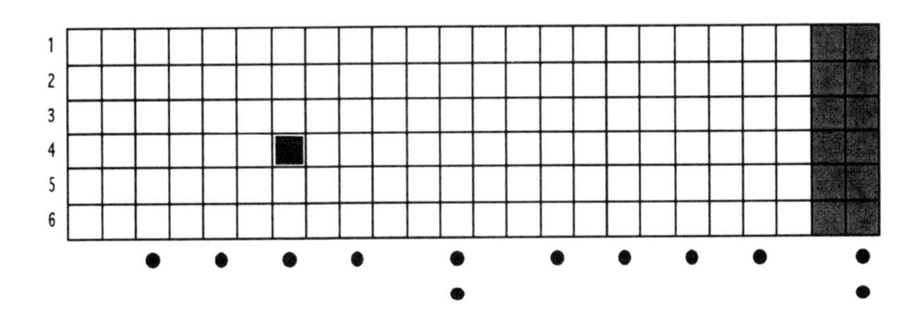

FIGURE 7.6. A graphic representation of a guitar fretboard.

Each display is associated with its own mode of operation. In normal operating mode, notes played on a MIDI guitar are seen on the active display. The notes are processed according to the translation modules seen on the function display, and the results affect the channels seen on the channel display. In translator selection mode, notes played on the guitar are seen on the function display, and are interpreted as switches that turn associated translation modules on or off. In channel selection mode, notes played on the guitar are seen on the channel display, and are interpreted as switches that turn channel assignments on or off. Both translator selection mode and channel selection mode are entered by using dedicated footswitches. Whenever a mode is entered, its associated display is highlighted.

The current version of the gesture translation software is written in Hyperlisp, an extension to Macintosh Common Lisp [2] [3]. The software is being used to experiment with realtime orchestration of music. A piece of music is pre-recorded using a sequencer. As the sequencer plays the music back, a performer uses available translation modules to alter various characteristics of each of the voices playing. Twenty-two translation modules are available; each can be associated with any string of a MIDI guitar. Nine of the modules alter sounds via program change messages. The remaining modules alter other characteristics of sounds, including reverb and chorus depth, pitch bend, volume, and modulation.

One translation module that is worthy of mention in more detail is the macro translation module. The current system allows the musician to group sequences of events into composite events called macros. Each macro thus created can be assigned its own notes on the guitar. When the macro translation module is active, playing a single note will cause all events grouped in the associated macro to take place. In this way, the musician can further simplify the sound-shaping process.

The creation of macros and their later use in performance can be seen as extensions of two modes of work familiar to every musician: rehearsal and performance. In rehearsal, a musician becomes familiar with a piece of music, decides ways to play particular passages in order to bring out nuances, and practices the gestures needed to play the piece until they become second nature. In performance, the musician focuses on getting the music across to an audience, using the results of rehearsal as a foundation. Similarly, a musician using the gesture translation software can experiment with various combinations of translation modules until a combination is found that works well in a particular musical passage. This combination can then be codified in a macro. In performance, the musician can concentrate on the music, using the predetermined macros as a foundation.

7.5 An Example

The gesture translation software was used to provide an arrangement of the first promenade of Moussorgsky's Pictures at an Exhibition. The music for the promenade was transcribed from the original piano score to a sequencer, with the melody, the bass line, and each of the inner voices given their own MIDI channel (six channels were used in all). Figures 7.7 and 7.8 show the music for the melody, annotated with gesture translation instructions. The instructions are in tablature, familiar to most guitarists. Each line of tablature represents a string, and the numbers on the lines represent fret positions on

FIGURE 7.7. Melody and gesture translation instructions for the first Promenade from Pictures at an Exhibition.

FIGURE 7.8. Continuation of melody and gesture translation instructions.

the strings. Numbers with boxes around them represent translation module selections.

A total of eight translation modules are used for the piece. The modules selected from frets 1, 2, and 7 are used for changing sounds on the Sound Canvas. Module 1 contains sounds associated with synthesizers, module 2 contains bass sounds, and module 7 contains percussion sounds. Module 17 controls vibrato depth, modules 18 and 20 control envelope attack and sustain rates, and module 21 controls resonance level. Finally, module 14 is used to select macros.

Initially, strings 1 to 6 (one being the highest pitched string) are set to MIDI channels 1 to 6 respectively, and each string is set to translation module 1. The channel settings remain constant for the duration of the piece. Before the music begins, translation module 14 is toggled on (turning on macro selection), translation module 1 is toggled off, and macro #1 (associated with fret 1) is selected. This macro is shown in Figure 7.9A. After the macro has been selected,

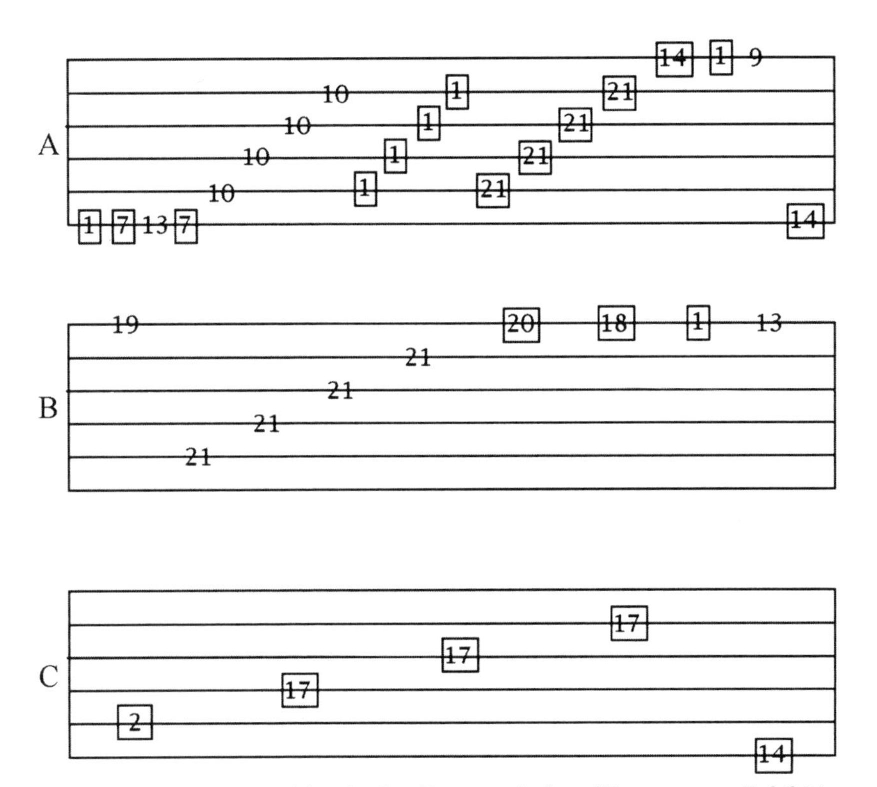

FIGURE 7.9. Macros used for the first Promenade from Pictures at an Exhibition.

string 1 has turned off translation module 14, turned on translation module 1, and selected a sound; strings 2 through 5 have selected a sound from translation module 1, turned it of, and turned translation module 21 on; and string 6 has turned off translation module 1, turned on translation module 7, selected a sound, turned off translation module 7, and turned on translation module 14.

The settings of macro #1 remain in effect until the end of measure 9 in the music. During this passage, variation is given to the sound on channel 1 (the main melody) by switching between the sound associated with the 7th fret and that associated with the 9th fret of translation module 1.

At the end of the ninth measure, string 6 is used to select macro #2, associated with fret 5. This macro is shown in Figure 7.9B. After the macro is selected, string 1 has selected a sound from translation module 1, turned on translation modules 20 and 18, turned off translation module 1, and selected values associated with fret 13 for en v elope attack and sustain; strings 2 through 5 have selected the value associated with fret 21 for resonance.

Macro #2 is in effect for measures 10, 11, and 12, during which time new values are given for resonance (on channels 2 through 5) and envelope attack and sustain (on channel 1).

At the end of measure 12, string 6 is used to select macro #3, associated with fret 7 and shown in Figure 7.9C. After this macro is selected, strings 2 through 4 have turned on translation module 17, string 5 has turned on translation module 2, and string 6 has turned off translation module 14. This macro remains in effect until the end of the piece.

In measures 13 through 19, various sounds (as well as resonance values) are selected for channel 5 and various values for envelope attack and sustain are selected for channel 1. In measure 22, new values for vibrato rate and resonance are given to channels 2 through 4. The piece ends with alternating values for envelope attack and sustain given to channel 1.

7.6 Related Work

Current research in controlling digital instruments has concentrated on two primary bodies of techniques. The first, which might be called gesture replacement, demands that musicians trade the technical skills on their instruments for a new set of techniques designed for controlling parameters of sounds. The most basic form of gesture replacement is illustrated by the front panel of the Sound Canvas, but there are more sophisticated alternatives. Devices such as the Radio Drum [4] are extremely easy to use, allowing parameters to be controlled by a performer's movements. Ease of use is often coupled with limitations in flexibility; the Radio Drum is limited by associating control parameters with positions in three-dimensional space. A more fundamental limitation of gesture replacement is that by its very nature it is incompatible with techniques for conventional instruments. The Radio Drum, for example, seems aimed at non-musicians who wish to interact with musical performances, rather than at musicians who desire a greater degree of musical control.

A second body of techniques might be termed gesture enhancement. Here, the musician uses traditional techniques on modified versions of traditional instruments. The modifications allow the musician's gestures not only to have their usual effect, but also simultaneously to control some set of sound-shaping parameters. The work with Hyperinstruments [5] is concerned with gesture enhancement. In contrast to gesture replacement, these techniques by their very nature allow compatibility with conventional instruments. The main drawback of gesture enhancement is that it can be inflexible. Parameters to be adjusted are tied to specific gestures. Since these gestures are also responsible for generating conventional sounds, isolation of parameter adjustments is difficult.

Additionally, in neither of the techniques mentioned above is there a correspondence to the rehearsal aspect of gesture translation as described in this paper. Using the techniques described here, a musician can custom-design a set of effects for a given performance. In other approaches, the effects to be used (and the gestures that activate them) are hard-wired into the system by those responsible for writing the software.

7.7 Limitations and Future Work

The current software is limited by its strong association with the guitar. While useful for guitarists, the software as it stands is of limited utility to other musicians. One direction for future work, therefore, is to broaden the ideas to apply to a larger class of musical instruments.

A second limitation relates to recovering from mistakes during performances. It is not unheard of for a performer to make a mistake during a performance. Further, given current technology for converting analog signals (e.g., notes played on a guitar) to digital information, it is not uncommon for a note to be incorrectly converted. When using macros involving many steps, small mistakes can have devastating effects. Therefore, another direction for future work is to investigate ways of safeguarding against catastrophes caused by either human or mechanical error.

Finally, the work presented here deals with only one kind of control,that of a specific sound-generating device. Future work will deal with allowing control over a broader range of devices, as well as control over events relating to the notes themselves. Such control could take several forms, including being able to alter the tempo at which a piece is being played, the ability to select the order in which parts of a piece are played, and the ability able to specify processes that generate new musical events in response to what is being played.

7.8 Conclusion

The gesture translation approach is intended to allow musicians to use existing techniques on their instruments in unconventional ways, to control parameters of digital instruments. When using this approach, a musician need not learn a complete set of new gestures for new forms of control; instead, he or she merely needs to learn a mapping between new forms of control and existing techniques. Additionally, these techniques can be completely freed from their conventional uses, and can be used instead for any form of control desired.

7.9 References

[1] Roland Corporation (1991). Owner's Manual: Sound Canvas MIDI Sound Generator SC-55.

[2] Chung, J. (1991). *Hyperlisp Reference Manual.* MIT Media Laboratory.

[3] Apple Computer, Inc.(1992). Macintosh Common Lisp Reference. Apple Developer Technical Publications.

[4] Matthews, M. V. (1989).The Radio Drum as a Synthesizer Controller. In *Proceedings: 1989 International Computer Music Conference.* Wells, T. and Butler, D.(Eds.), San Francisco: Computer Music Association.

[5] Machover, T., and Chung, J. (1989). Hyperinstruments: Musically Intelligent and Interactive Performance and Creativity Systems. In *Proceedings: 1989 International Computer Music Conference.* Wells,T. and Butler,D.(Eds.), San Francisco: Computer Music Association.

8

Visualization for Personal Information Systems

Xia Lin[1]

8.1 Introduction

Everyone has a personal information collection that includes various documents such as books, journals, article reprints, notes, and memos. Every researcher spends a large amount of time collecting and organizing information for his or her personal collection. With the advance of computer technologies, larger and larger portions of personal collections are stored in computers. However, searching information in one's computer is quite different from finding a book or paper in one's office: while books and papers are visible and recognizable, information in the computer is not.

Because of unique characteristics of personal collections as discussed later in this paper, designing and implementing information systems that can help the user to effectively organize and use personal collections has been a great challenge. This paper discusses one important aspect of personal information systems, that is, the need to have a visual space for personal collections. In this paper, characteristics of personal collections are first summarized. Issues related to personal information systems are reviewed. Attention is called to the fact that much useful visual and spatial information is lost when personal collections are stored in computers. To make up the loss, or at least a portion of it, a personal knowledge space is proposed for personal information systems. The knowledge space is a mapping of the high-dimensional information space onto a two-dimensional space. It defines a framework for a personal collection, and shows interrelationships of items in the collection through various graphical features. It provides a visual space that one can interact and become

[1]School of Library Science, University of Kentucky, Lexington, KY 40506.

familiar with. Using an example of such a knowledge space that we mapped from a researcher's personal collection, we discuss the characteristics of the space and various potentials of using the space for personal information systems. Finally, future research is proposed.

8.2 Personal Information Systems

Personal collections provide a unique contribution to one's information needs. Because of the personalized arrangement, the convenient location and the flexibility to meet and reflect individual needs and interests, personal collections are more accessible than any other collections of materials [1]. Furthermore, materials in personal collections are generally available when one needs them; they can be recognized by their appearance because they are familiar to the user; and they contain personal notes and ideas that help to integrate one's thought collected at various instances throughout ones major activities. All these unique features of personal collections make system design for these collections difficult.

People collect, annotate and store personal collections according to their own idiosyncratic needs and preferences. The content and organization of their personal collections often reflect the nature of work, the method of working, the technical environment and the personality of their users [2]. Thus, their desires for personal information systems are not simply a database or a retrieval system that contains their personal collections, but a tool "that enables better usage of accumulated information, that stimulates creative thinking, and that improves your style of intellectual work" [3]. Specifically, Stibic defined functions of a personal information system as:

- Preventing information once gained from being lost,
- intensifying the use of available information resources,
- improving the organization of knowledge,
- offering a superb opportunity for creative use of information for easier linking of factors and ideas, and
- uncovering unseen relations, associations, and conclusions.

Previous approaches for personal information systems date back to Bush's famous Memex. "A Memex is a device in which an individual stores all his books, records, and communications, and which is mechanized so that it may be consulted with exceeding speed and flexibility. It is an enlarged intimate supplement to his memory" ([4], p. 106). Bush's idea was to build Memex as a personal tool to store and associate information in one's personal collection in a form close to the human mind. He emphasized that links and associations needed to be developed in personal information systems.

Since then, many researchers have designed and experimented with various personal information systems. On the high-end, these researchers looked for systems that would augment man's intellect and share with man in the problem-solving process [5][6][7]. On the low-end, they designed systems that could be utilized for personal information organizing such as indexing and bibliographic control [8][9]. While these systems emphasized various aspects of personal collections, we found that the visibility of personal collections in one's working environment, an important aspect of personal collections, was missing from most of these systems.

Empirical studies show that an important function of personal collections is to remind people of what needed to be done and what is available [10]. People often find things in their personal collections because they can recognize them when they see them. They associate materials in their personal collections not only by information context, but also by physical context, emotional context, and time or event related clues [11]. Through repeated use and reuse of materials in their personal collections, they establish a mental image about the content, the organization, and spatial locations of the collection. Georgia Miller ([12], page 287) once had a vivid description about this:

> Every working day I sit in an office at a desk cluttered with piles of papers in a room whose walls are covered with books and journals. After a few months in that room I had learned where things were. I feel in my muscles that psychological journals are here, and linguistic journals there, and computer publications over there. I know what is in each pile on the desk, and I have a rough idea which file hold which papers. In short, I work in a cubicle that is lined with information which is spatially represented in my memory.

When materials are stored in computers, they become invisible to people. All clues for visual perception and associations disappear. Those once very helpful visual cues no longer exist, such as the color and the art design of a book cover, the look of the first page of paper reprints, or highlighted marks on the papers. What also disappears is the spatial information that one is used to, such as the office layout, the different locations of drawers, cabinets and bookshelves, and the piles and files on the desk. Access to personal collections stored in computers often depends on the searchability of short file names that represent materials in the collections. Without any visual help, one may only get familiar with materials by their contents, not by their appearances and locations. As a result, there becomes less and less a distinction between materials in one's personal collection and other materials one can get through computers. (A typical example is that all articles printed from an online full text database look similar, despite their varieties of sources.)

It is clear that we need to restore visual elements when we store personal collections in computers. We need to make information in the computer visualizable to people, and we need to make it easier for people to recall and recognize the information when they see it. The computer has the advantages of fast processing speed and quick matching capability. The human being possesses a unique perception and pattern recognition capability. Merging the advantages of both the computer and the human being into the personal information environment should be one of the most important considerations for designing visualizable personal information systems.

Our approach to such personal information systems is to create a visual space by which one can view contents, locations, and relationships of materials in one's personal collection. In the following, we will propose a map display to visualize a personal collection. The map display will show the underlying structure of the personal collection and display the contents based on inter-relationships of items in the collection. The map display will be generated automatically based on a neural network's learning algorithm, Kohonen's feature map. Details of the mapping algorithm and its applications to the document space was described in Lin et al. [13] and Lin [14]. The next section discusses a result of applying the algorithm to a personal collection to generate a map that we call personal knowledge space.

8.3 Personal Knowledge Space

A personal knowledge space can be defined as a map of a person's subject areas represented in his personal collection. Ideally, the map should cover all major areas that the person has knowledge of or is interested in. The map should show a clear distinction between major and minor subject areas of the person, and it should be organized to reflect relationships of the subject areas from the point of view of that person. The map should be a guide to the personal collection; one should be able to use the map to show what is in his personal collection, just as one can use a street map of a city to describe attraction areas in the city.

An experiment was conducted to generate such a map using Kohonen's feature map learning algorithm. The algorithm was selected because it offers the possibility to create in an unsupervised process topographical representations of semantic, non-metric relationships implicit in linguistic data [15]. The algorithm is a mapping process that maps a high-dimensional space onto a two-dimensional grid. It preserves the most important relationships among the input data and makes such relationships geographically explicit. It creates an economic representation of the high-dimensional input data on the two-dimensional map by allocating different sizes of domains to different inputs based on their occurrence frequencies and their interrelationships [16].

A researcher's personal collection were collected in a HyperCard stack. The collection contained 660 documents, which were accumulated over many years as a by-product of the person's research activities. Each document in this collection contains a citation (author, title, source, etc.), an abstract and sometimes the first page of the article. We first did a preprocess that included the following four steps:

1. identification of unique words in the collection after necessary truncations,
2. exclusion of the stop words,
3. frequency counts of the words identified, and
4. exclusion of the most and the least frequently occurring words.

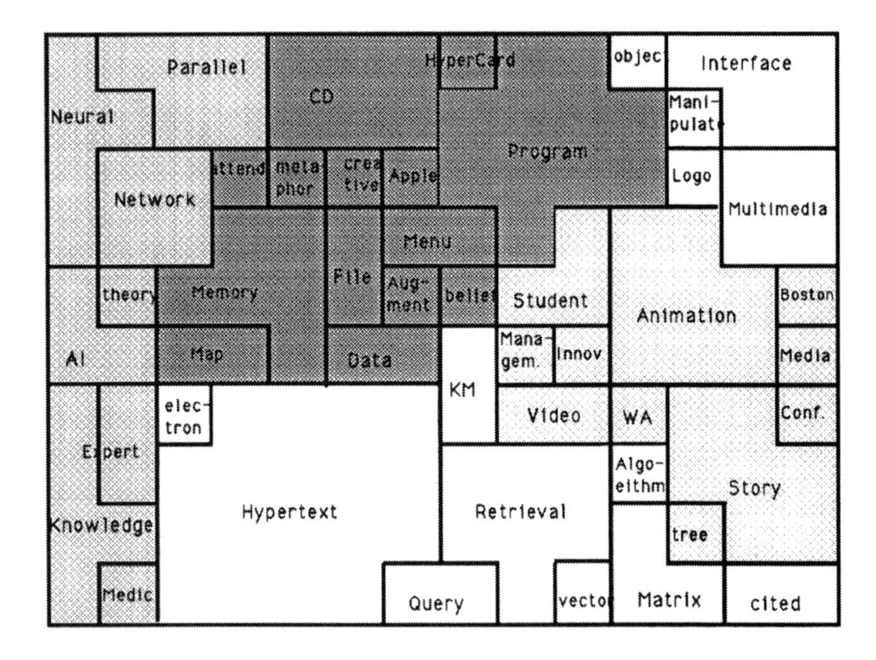

FIGURE 8.1. A map display for a personal collection.
The display shows the person's major research areas and their relationships. The size of areas indicates relative importance of the areas to the researcher. The neighboring relationships of the areas reflect associations of these areas as derived from the collection.

The preprocessing resulted in 1472 unique words which were then used to index the collection, creating 660 vectors of 1472 dimensions. After a normalization process, these vectors were used as input to train a 10 by 14 Kohonen's feature map. Figure 8.1 shows the result after 2500 learning iterations. The whole learning process took about 202 seconds in a Cray super-computer.

As the result shows, the mapping of such a personal collection creates a personal knowledge space. The space shows the researcher's major research areas and the relationships of these areas. The size of areas, corresponding to the frequencies of the words, indicates relative importance of the areas to the researcher (the more often a word appears in the personal collection, the more likely the word will correspond to a large area in the space). The neighboring relationships, corresponding to the frequencies of co-occurring words, reflect associations of the areas resulting from the activities of the researcher. Some of these associations, such as the association of query with retrieval, are clear to anyone in the field of research, while other associations may need the interpretation of the researcher. For example, without knowing that the researcher was interested in animation for story-telling to students, we would not understand why story and student were represented as two major areas near the area marked animation.

The space appears to have a hierarchical form. Four top levels of the hierarchy can be identified as General AI, General computing and programming, Hypertext and retrieval, and Multimedia and animation (indicating on the map by the colors or gray areas). Within each level there are subject areas organized by their inter-relationships. For example, within the first hierarchical level on the left, traditional AI and neural networks are located on different side of the area AI. This indicates two different but associated approaches to artificial intelligence; while parallel is close to neural networks, expert (systems) is close to Knowledge (base), and so on. When we looked further into the neural networks area (through a more thorough analysis of mapping results not shown on the map), we found that words mapped to this area were: connectionist, model, learning, process, unsupervised, net, error, and principal component analysis (Evidently, the researcher is interested in comparing neural networks to principal component analysis.) Similarly, words mapped into hypertext area included link, text, color, structure, browsing, and University of North Carolina at Chapel Hill (where the first hypertext conference was held).

Each document in the collection also has a unique position in the personal knowledge space. These document positions show the distribution of the personal collection over the personal knowledge space. This distribution may remind the researcher of the areas on which he has more documents and of the areas on which he may need to have more documents. The distribution may also be used to show migration of the researcher's interest over time. Figure 8.2 gives the distributions of the first 100 documents and the last 100 documents in the collection (these maps are the same map as figure 1 except the selected documents are represented as dots on the display, no new mapping is needed). Since the documents are collected sequentially over time, we can see a clear migration of the researcher's interest from early more broad distribution over several

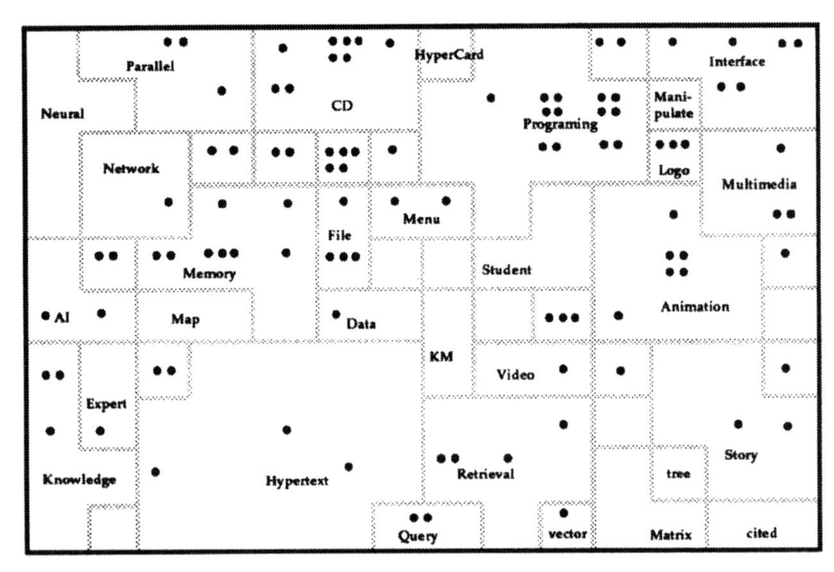

(a) Distribution of the first 100 documents in the personal collection

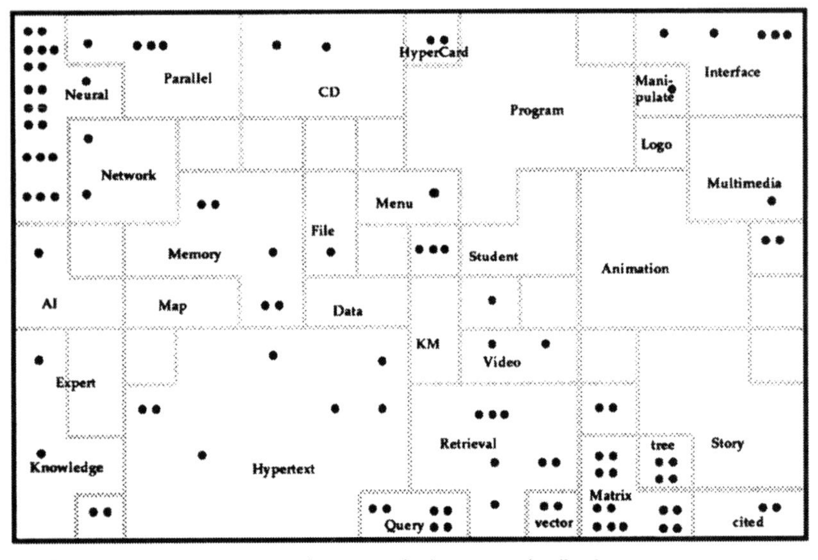

(b) Distribution of the latest 100 documents in the personal collection

FIGURE 8.2. Document distribution on the map display.
The two different time periods of documents showed very different distributions, including the change of the personal research activities.

areas, such as CD-ROMs, programming languages, and interfaces, to a more current concentration on neural networks, matrix operation and retrieval. The interpretation, of course, will make more sense to the researcher than to others. When the map was presented to the researcher, he confirmed that those areas and relationships shown on the map were basically correct. The researcher found that what we interpreted from the map sometimes even revealed interesting relationships that had not been considered before. He thought that the map was a reasonable picture of the total corpus of the collection, but not necessarily of his current status (this is because the map was generated based on his past six years collection, not just on his more recent collection). He suggested that in a real system of this sort, there might need to be a way to age and give less weight to documents which were old or hadn't been accessed in a while.

8.4 Discussion

There are many advantages of the knowledge space, including automatic abstracting of key words from the data, automatic labeling of the displays, and automatic proportioning of the displays based on statistics of the input data. Another obvious advantage of this two-dimensional knowledge space is that it can be showed on the computer screen. This makes it possible to use the map display as an interface for a personal information system. A prototype system in HyperCard was designed to experiment with potentials of the map interface for retrieval systems [13]. The prototype system was designed in HyperCard using the map display as an interface (Figure 8.3). The map interface provides an overview to the materials in the system and provides implicit links among related documents. Such links, when used together with other links provided by the system such as word links, item links, index links, and associative links, become an efficient navigation tool for the user to explore and access materials in the system.

The map interface can support functions that are particularly desirable in the personal information environment. In the following, we will discuss several of such functions based on the personal knowledge space and the prototype system.

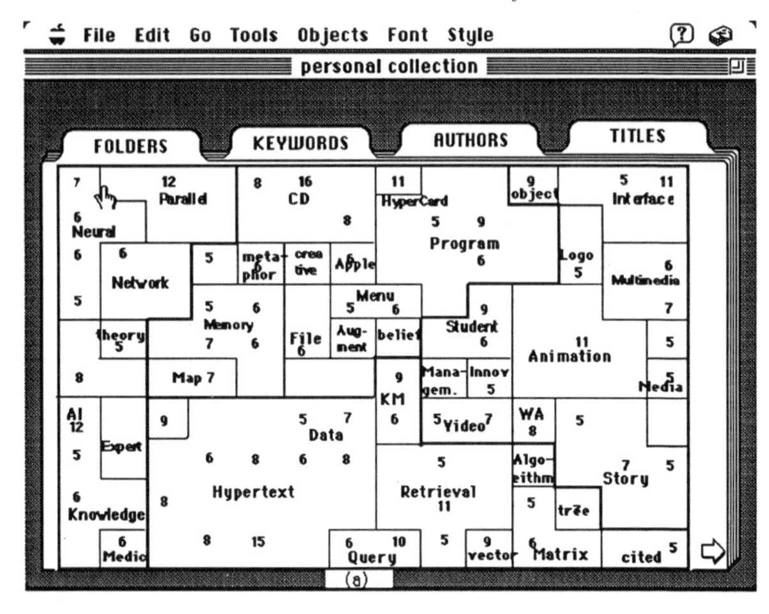

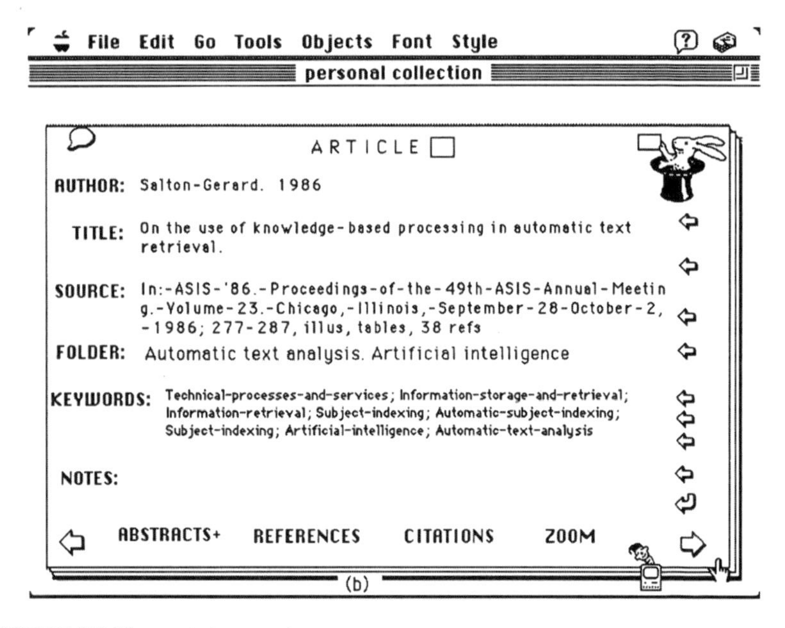

FIGURE 8.3 The prototype system.
a) The map interface. The map shows major areas of the collection as well as the document distribution over the content map. Each number on the map represents a node on the map, indicating the number of documents mapped to the node. For clarity, nodes that have fewer than five documents are omitted from the display.
b) A full record card. Each record card contins three types of linking buttons: the left button (author,title,etc.) are the link buttons; the right column (arrows) are the item link buttons; and the bottom row are associative link buttons. Selecting any of the associative links will pop up a window to show the corresponding contents.

8.4.1 Visual Familiarity

As discussed earlier, materials in one's personal collection were generally familiar to the user, both in terms of their contents and their locations. People will often be reminded of their past experience when they are prompted with familiar materials in their personal collections. Using the map display as the interface for a personal information system will create a visual space that the user can explore and become familiar with. Since the map display is organized by contents and associations of one's personal collection, it helps to remind the user where a document is located on the display, particularly when the user becomes visually familiar with the spatial arrangement of the display.

The issue of visual familiarity is very important for systems created for a personal information environment. We all have the experience of being lost in a large complex building. A building interior structure can look very complex for a visitor, but it seldom creates orientation problems for people who work there daily. This is because people become familiar with the building and establish a visual image of the building layout. Similarly, the map display may look complex and incomprehensible for people who have not become familiar with it, but it will look and feel highly organized, psychologically and personally, to the user who utilizes it often. The clusters, the areas, and various visual cues on the map, will have special meanings to the user, just as folders, drawers, and various organizing means in one's personal collection will have special meanings to oneself (By special we mean, for example, a word network on one's folder will mean much more than its literal meaning; it can remind one that this is the folder for papers from a conference on internet networking, or this is the folder for all the correspondence to people in one's professional network, depending on where the folder is located in one's office).

8.4.2 Links by Associations

Another function that the map display supports is linking by associations. As demonstrated in Memex and other experimental systems, linking is a major function desirable in the personal information environment to support browsing. The map display creates semantic links by organizing related documents together. Such semantic links have two advantages. First, the links are generated automatically through the learning process based on information in the collection. Second, the links are implicit, they are shown through item locations on the display and the relationships among the subject areas. Because of this, one may recognize different links from the same display when one has different needs in mind.

The map display also provides an overview of the underlying collection. Since the overview is organized according to inter-relationships of items in the collection, browsing through the display will become much easier. The user can start from the map display, select to view one or several clusters or areas, then follow various links (such as word links, item links, indexing links and associative links supported by the prototype system) to get to an item. When a necessary item is found, the user can go back to the map display, examine the location of the item, and select other items near the item. Because of the associative arrangement, these items are likely relevant to the necessary item.

8.4.3 Filters and Monitors

People collect documents in their personal collections based on their personal needs and interests. Thus, the personal collection reflects these personal needs and interests. Once we have a map for the personal collection, we have a map of the person's interests. This creates the possibility to use the map as a signature of the person' s interests for monitoring materials that may be interesting to the person. The way the map display is produced made such a monitor process easier. Since not only the display is generated, the underlying network whose weight patterns is tuned by the input patterns (the personal collection) is also constructed during the mapping process. The network thus can be used as a filter to test if other materials are related to the topics represented by documents in the personal collection. New information items whose weight patterns are similar to those input patterns can be retained on the network and represented visually on the display. The visual patterns of the new information items will then prompt to the viewer about its contents, and its relationships with those items in the personal collection, allowing the viewer to decide whether the new items are of interest to him or her.

In the near future, the whole world of information will be connected. Most of the information will be accessible from computers if the user knows when and where the information needed is available. While no one will constantly monitor all information resources on the network, a personal information agent using the map display as a filter can be designed to routinely look for needed information in various information resources on the network. Assuming that we can let the information agent work automatically throughout the night, we will have a monitor that can routinely keep track of new materials in online databases and networks. Every morning, when a researcher comes to her office, she will notice new documents by scanning through the map display, or her personal knowledge space. She will become used to those patterns representing documents that she is likely interested in. Thus she can decide if she should read more about the new documents and whether they should be added to her personal collection or be discarded. When the new documents are kept with the personal collection, they will eventually change the map display and the monitor through updating and re-training the network later.

8.5 Summary and Future Research

A personal knowledge space is desirable for an effective personal information system. Such a personal knowledge space should restore visual elements of personal collections. It should create a semantic space that one can explore and become familiar with. It should provide a filter and a monitor to let one keep track of the world of electronic information. By using Kohonen's feature map, we constructed an example of such a personal knowledge space to demonstrate the desired potentials. Using such a knowledge space as the interface for personal information systems, we constructed a prototype system that will let users interact with their personal collections through association links and browsing, as well as establishing visual familiarity.

Currently, we are continuing to study other features that the personal knowledge space demonstrated. We have conducted two experiments to examine validity and usefulness of the structure shown on the knowledge space. One experiment was to compare such a space to similar spaces generated by human subjects and by other automatic algorithms. The other experiment was to let some human subjects perform certain retrieval and browsing tasks on the knowledge space. Results of the two experiments are reported in Lin [14]. We are also comparing the knowledge space with geographical maps. The graphical representation of physical space should provide insights to the graphical representation of information space. In addition, we are also studying cognitive maps that subjects draw to represent their personal collections. We plan to ask subjects to recall what they have in their collections and to organize them in a meaningful way on a map. These studies will lead to a better understanding of how we mentally categorize, organize, and represent personal collections, and will eventually help to define how the personal knowledge space should be structured.

The representation on the personal knowledge space should be enhanced by more creative graphical assistance. Such assistance may include use of colors for different subject areas, use of icons to indicate different type of documents, having hierarchical sub-maps so that more detail relationships can be examined, establishing conventions to represent certain semantic links, etc. With future information technologies, we can explore the possibility of having a wall-size knowledge map on one's office. Instead of covering the wall with books, journals, and other documents, we can have a map on the wall that is well organized to the needs of our work, that is a familiar space in which we can feel what is where, that can help us keep track of the world of electronic information, that has various links and trails to let us freely travel around, and that can let us zoom in and out to any chunk of information, from a book, to an article, or to just a section of notes.

8.6 References

[1] Soper, M. E. (1976). Characteristics and use of personal collections. *Library Quarterly*, 46(4), 397-415.

[2] Leggate, P. et al. (1977). An On-Line System for Handling Personal Data Bases on a PDP 11/20 Minicomputer. *Aslib Proceedings*, 29 (2), 56-66.

[3] Stibic, V. (1980). *Professional Documentation for Professionals: Means and Methods.* Amsterdam: North-Holland.

[4] Bush, V. (1945). As we may think. *Atlantic Monthly*, 176(1), 101-108.

[5] Engelbart, D.C. (1963). A conceptual framework for the augmentation of man's intellect.

[6] Kay, A; Goldberg, A. (1977). Personal Dynamic Media. *Computer,* (March), 31-44.

[7] Card, S.K, Moran, T.P. (1986). User technology: From pointing to pondering. In *Proceeding of ACM Conference on the History of Personal Workstation.* New York: ACM, pp. 183-198.

[8] Burton, H.D. (1981). FAMULUS revisited: ten years of personal information systems. *Journal of the American Society for Information Science*, 32(5), 440-443.

[9] Heeks, R. (1986). *Personal Bibliographic Indexes and Their Computerization.* New York: Taylor Graham.

[10] Malone, T.W. (1983). How do people organize their desk? Implications for the design of office information systems. *ACM Transactions on Office Information Systems,* 1(1), 99-112.

[11] Christie, B. (1986). Personal Information Systems. In B. Christie (ed.), *Human factors of information technology in the office.* London, pp. 127-144.

[12] Miller, G.A. (1968). Psychology and information. *American Documentation*, 19, 286-289.

[13] Lin, X.; Soergel, D; & Marchionini, G. (1991). A self-organizing semantic map for information retrieval. *Proceedings of the Fourteenth Annual International ACM/SIGIR Conference on Research and Development in Information Retrieval* (Oct. 13-16, 1991), Chicago, Illinois, pp. 262-269.

[14] Lin, X. (1993). *Self-organizing Semantic Maps as Interfaces for Information Retrieval.* Doctoral Dissertation, College of Library and Information Services, the University of Maryland at College Park.

[15] Ritter, H. & Kohonen, T. (1989). Self-organizing semantic maps. *Biological Cybernetics*, 61, 241- 254.

[16] Kohonen, T. (1989). *Self-organization and Associate Memory, Third ed.* Berlin: Springer-Verlag.

9

Blazon Computer Graphic Notation

Judson Rosebush[1]

9.1 Overview and Background

A herald is a graphical symbol for an individual or organization. In the simplest case, a herald may be a field of a single color. In practice, the visual field is usually divided into two or three colors and may also involve iconographic representations of animals, plants and objects. Modern heraldry evolved in Europe, particularly in England and the Romance Countries. The subject has a rich literature; the references cited are typical contemporary overviews and contain many further references.

At first, heralds were probably just adopted. But as commerce and trade grew, heralds became regulated, and duplicates were discouraged. In 1407 the King of France incorporated a College of Arms to log drawings as well as formal written descriptions of individual heralds into the public records. This necessitated the written descriptions, called blazon, become standardized so that a herald could be unambiguously recorded and reconstructed from only the written description. In modern terms the written description can be completely represented with a subset of ASCII characters.

The rules of blazon provide a way to parse textual commands so as to define a graphical field, partition it into sections, place lines on it, define objects within it, and create repeating forms. Blazon accomplishes this task by defining a very primitive set of graphical operations and rules for their use. As a system, blazon is pre-Cartesian, that is, it does not involve the use of a quantitative X and Y axis coordinate geometry. It is largely qualitative in description, although some functions do provide a capability to incorporate integer ar-

[1]Judson Rosebush Company, 154 W. 57th Street, No. 826, NY, NY 10019

guments. The language of blazon reveals insights into fundamental low-level graphical operations. Furthermore, its syntax also incorporates inflections of language, whereby primitive functions may be "declinated" to compress syntax in a matter akin to adverbs and adjectives in normal language.

One purpose of heraldry is to graphically depict the genetic composition of an individual; a special processor called the Marshalling of Arms recognizes two parent shields as inputs and formulates an output composite off-spring shield as a function of the two inputs. In practice, marshalling occurs once per generation, with the output shield becoming increasingly fragmented, yet still a visualization of the genetic encoding of DNA. In this regard heraldry is a scientific visualization.

The purpose of this paper is not to describe heraldry, blazon, or marshalling in detail-rich literature already exists in this field. Our motives are to introduce the subject in a computer graphics context. Blazon may be the oldest computer graphics language, nevertheless it is by no means extinct. A study of heraldry and blazon represent an archaeological dig into the early history of computer graphics, one which incorporates 600 year old executable computer graphic notation. The notation still works and is useful today.

Secondly, the fundamental blazon syntax reveals significant wisdom into computer graphic languages in at least two major ways. First, its fundamental operators yield insight into primitive language design. Secondly, its striking capability to declinate operators may well be an overlooked yet fundamental concept that can be applied in contemporary language design. We will examine both of these in detail.

9.2 Basic Terms and Definitions

The modern reader needs to understand the fundamental terms and operations of blazon. The following descriptions attempt to cast these definitions in terms of a modern, notational computer language.

9.2.1 The Field and the Points of the Field

The field, also called the ground, the escutcheon, or the field of view, is akin to what in computer graphics is commonly called a window, except it is usually shield-shaped. There are a few different field shapes. The most common is the shield shape, such as what is shown in the drawings here, but the reader should be aware that there are many variations in the shape of the shield, especially with regard to its ratio of width to height, and with regard to the curves that form the bottom edge. In addition to the shield shape a diamond shape is sometimes used to denote the female sex.

The field is addressed qualitatively by names which refer to its top, bottom, left, right, and two center axes (Figure 9.2). These divide the shield into nine points, which are the various regions of the shield. The top of the shield is called the chief, the bottom the base, and the middle the fess. The right of the shield is called the dexter side, the left is called the sinister side and the middle the pale.

To be right handed is to be dexterous, lefthandedness in French is gauche (and thus bad, sinister). But note that the dexter and sinister sides are relative to the holder or bearer of the shield, who is presumed to be facing you, the onlooker or the beholder. Thus the dexter side of the shield is the right side of the person holding the shield, as if behind it, and thus it appears to you, the beholder, on your left. Thus the notation is sophisticated enough so as to distinguish the shield independently of the viewing environment, an analogy similar to the concept of stage-right or stage-left in the theater.

9.2.2 The Plain Shield: Tinctures, Metals, Furs, Tricks

The surface of the field is one of three different materials– a tincture, a metal, or a fur. There are at least five tinctures (azure, gules, sable, vert, purpure), two metals (or, argent), and four furs (ermine, erminois, pean, vair). The tinctures are colors (blue, red, black, green, purple), the metals are metals (gold, silver), and the furs are animal fur (ermine, erminois, pean, vair). Conventions exist about which of these materials may be combined together onto a single shield.

In 1639 Marcus Vulson de la Colombiere standardized a system of black and white cross-hatching to represent the different materials of a heraldic shield. This remains in use and is illustrated here (Figure 9.3). In heraldry, the drawing of an outline of a shield with cross-hatching to indicate color is called a trick; the verb "to trick" is the process of creating such a graphic.

The most basic shield is of only one material, e.g., the plain azure shield of Berington of Chester (Figure 9.1). When a shield consists of only one material, the blazon of the shield is the material. Obviously there are only a handful of such shields.

9.2.3 Partitioning the Shield and the Ordinaries

Given the 11 materials described above it is only possible for 11 families to possess solid shields, and Fox-Davies claims only one of these histories exist. In most shields the field consists of more then one material, and thus involves a division of the field.

The simplest way to divide the shield is to employ a single partitioning axis, dividing the field into two regions, either horizontally (fess), vertically (pale), or on the right or left diagonal (bend, bend sinister) (Figure 9.4). Two partitioning axes create a cross, a saltire, a chevron, or a pile (also Figure 9.4). These partition axes, as well as the resulting forms, are called ordinaries, and the process, or action of partition is said to be a party.

When writing blazon the verb party is followed by a qualifier, or preposition, which describes a particular type of party, followed by the name of the ordinary. Two common prepositions are per and a; "per" indicates a division into two regions, and "a" indicates that a stripe is placed at the division axis, dividing the shield into thirds. Thus the various ordinaries may be described in terms of their party, e.g., party per fess or party per bend, although the verb "party" is some times elided (per fess, per bend). A complete blazon obviously incorporates the materials names as well. The ground material, that which makes up the background of the shield, is listed first and before the preposition and ordinary, and the foreground material is listed last, for example, gules party per fess argent (see Figure 9.1). Again, often the word "party" is elided, and the blazon would simply be gules per fess argent, or per fess gules and argent.

9.2.4 The Thirds Ordinaries

A number of ordinaries are also based on dividing the shield into thirds. Many of these are related to the ordinaries already described, and are blazoned by stating first the background color, then the partition type (in this case, "a"), the ordinary, and finally the color overlaid onto the new shape, as in azure a pale or (Figure 9.5).

9.2.5 Sub-Ordinaries and Discrete Charges

Beyond the ordinaries already described there also exist a variety of other standard definitions to assist in partitioning the field. Sometimes these are also called ordinaries; other writers call them sub-ordinaries. Several of these result from amalgamating different points of the shield (Figure 9.6). The canton is the dexter chief point (see Figure 9.2), or top left corner of the shield. The chief is the top third of the shield. A quarter is the upper one quarter of the shield. A pall is a Y-shaped figure.

Geometrically shaped discrete charges may also be placed on the field (Figure 9.7). The inescutcheon is a small figure shaped like a shield in the middle of the shield, and used to denote the arms of a heraldic heiress. A billet is a rectangular shape. Thelozenge is a parallelogram having equal sides forming two acute and two obtuse angles. A mascle is a lozenge voided. The roundel is a circular form. The fret is a mascle interlaced with a saltire; it is symbolic of a weave and the diagonals change priority relative to the viewer. The fusil is a tall diamond. An annulet is a roundel voided. The rustre is a lozenge with a circle in the middle. Flaunches are two arches at the sides of the shield.

Discrete charges can also include animals, such as the lion, and here the language of blazon provides a vocabulary for different positions, e.g. vert per chef vair with loin rampant (Figure 9.8). Discrete change may be applied in abundance, and blazoned with the type of discrete change (used as a verb) preceded by the quantity. The background material is described first in the blazon and the foreground material is described last, for example, sable ten roundels or, or azure three billets or (Figure 9.9).

9.2.6 Repetition of Form

Repetition of form is specified in blazon and involves the formation of new nouns which describe the shield and which, by and large, derive from the root ordinary. The new noun is usually formed by the addition of a "-y" suffix to the root ordinary name. In the case of the ordinaries, this new noun is then followed with the phrase "of N" where N is the integer number of repeats, written as a word. A pale repeated becomes a paly, a bend a bendy, a chevron a chevronny, and a cross a chequy (Figure 9.10). A few others bear little resemblance: a fess repeated becomes a barry, and a bend sinister repeated is called a compony (again Figure 9.10). A gyronny divides the field into eight radial divisions. A barruly is a field with multiple bars. These nouns can also work as adjectives: a bend barry is a bend with horizontal stripes; a bend compony is a bend with diagonal stripes.

Discrete charges also acquire special names when they dominate the field. More than ten billets placed on the field become billett. Many lozenges become lozengy, an overall pattern of frets become fretty (Figure 9.11). Many crosses repeated on a field are called a crusilly.

9.2.7 Partition Lines, Diminutives, and Borders

Blazon provides syntax for different kinds of partition lines between the various tinctures (Figure 9.12). This extra description is incorporated in a blazon immediately following the partition line, for example, sable a fess engrailed or. Examples presented here include the fess (Figure 9.13a), the bend (Figure 9.13b), the pale (Figure 9.14a), the chevron (Figure 9.14b), the cross (Figure 9.15a), the saltire (Figure 9.15b), and the chief (Figure 9.16).

Ordinary or Charge	Diminutive	Repeat
fess		barry
bar	barrulet	barruly
bend	bendlet	bendy
bend sinister		compony
pale	pallette	paly
chevron	chevronel	chevronny
cross	crosslet; canton	chequy; crusilly
saltire		
pile		
billet		billetté
lozenge		lozengy
fret		fretty

Other blazon terms pertain specifically to modifying the width of charge. The diminutive of a charge is a charge with narrower width. This is often formed with the suffix "-let" to the ordinary name (above table). A bendlet is a diminutive of the bend and is a charge one-half the width of a bend, and a cottise is a charge one-half the width of a bendlet. A pallette is a diminutive of a pale and is one-half the width of a pale; a band one-quarter the width of a pale is said to be a pale endorsed. A bar is a narrower version of a fess and normally occurs in pairs; the diminutive of the bar is the barrulet. The diminutive of the chevron is the chevronel. The diminutive of the cross, where the cross is repeated fractal-like, is called a crosslet. The diminutive of the quarter is the canton. Several blazon terms pertain specifically to modifying borders. A bordure is a narrow inscribed border, that follows the shape of the shield. An orle is a border that is one-half the width of a bordure and typically lies inside it. A tressure is a pair of orle placed concentricity, also described as an orle gemel.

Finally, a few miscellaneous terms: An enhancement is a charge that is raised in position on the field. Gemel means to occur in pairs. Couped means that the charge does not bleed off the shield, as a cross couped, or a fess couped. Sem means "powdered," or strewn with minor charges. Gutte means strewn with minor charges which are drops of a liquid. Rompu implies that a charge is sub-divided by another charge.

9.3 The Marshalling of Arms

Finally, one must understand that blazon not only describes static artwork, but also involves a temporal, animation aspect. In this regard its vocabulary is limited, for it includes only one temporal operator which produces a discrete temporal sequence. That operator is called the marshalling of arms and involves the processing of two input shields and the production of a composite, resultant shield. The two input shields are those of a father and a mother, and the output shield is the shield of the offspring. In this regard the heraldic marshalling of arms processor is a true scientific visualization, in that it produces a symbolic, graphic visualization of the genetic of the subject.

The marshalling of arms processor is very simple and works like this: The resultant shield is quartered, and the shields of the two parents are reproduced (smaller) in opposite corners, with the father shield placed on the top left and bottom right, and the mother shield placed on the top right and bottom left. In successive generations, the process is repeated recursively (Figure 9.17).

9.4 Summary of Blazon

In summary then, the basic vocabulary of heraldry includes variable names for a) materials, b) ordinaries (the angle of partitioning lines), c) the partition operator (halves or thirds), and d) the shapes of edges of the partitioning lines. The basic syntax is backgroundmaterial operator ordinary linetype foregroundmaterial, and assembly uses a painter's algorithm. If linetype is elided it is assumed to be straight.

But beyond this there is a diminutive form of the ordinary, in which the ordinary noun is declinated to indicate a narrower version, as well as the repeated form of the ordinary, in which the ordinary noun is declinated to indicate pattern repeat. The language of blazon provides for a rich description of dividing the plane of the shield. Consider for example the following blazons, each of which apply to a bordure region: bordure per pale, bordure quarterly, bordure gyronny, bordure compony, bordure chequy.

Finally, the marshalling of arms is a graphical image processor which provides a scientific visualization of the genetic makeup of an individual.

9.5 References

[1] Fox-Davies, A. C. (1969). *A Complete Guide to Heraldry.* New York: Bonanza Books.

[2] von Volborth, Carl-Alexander (1981). *Heraldry: Customs, Rules and Styles.* Poole, Dorset, England: Blandford Press.

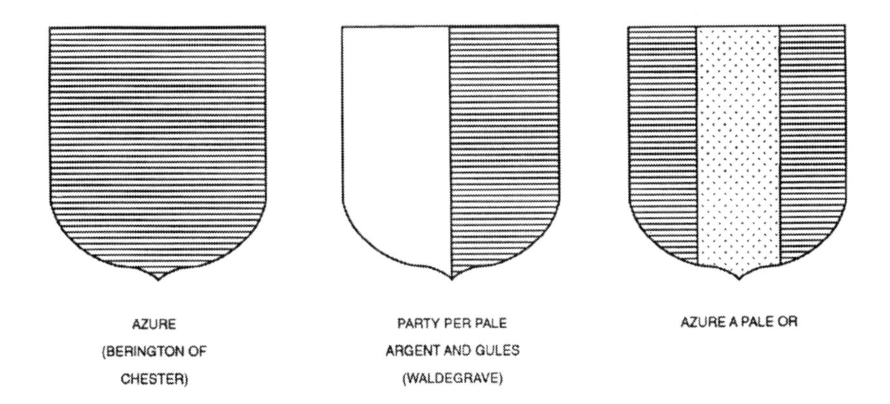

FIGURE 9.1. Three different heralds, and below them their blazon. In two cases the families who bear the shield are identified. (Drawing by Matt Alexander and Rosmarie Awad).

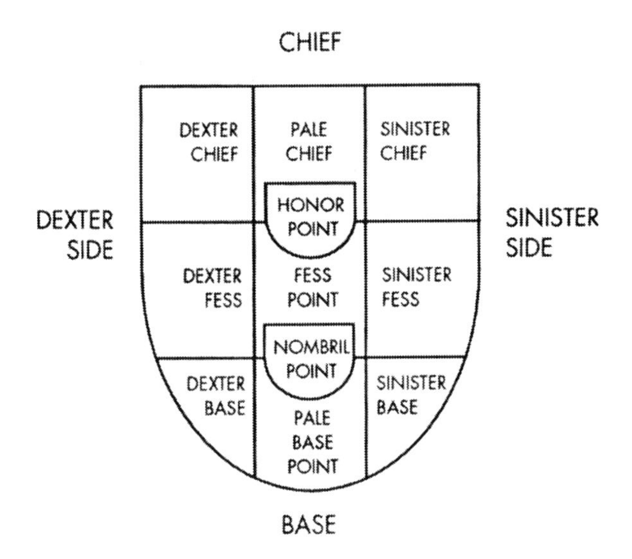

FIGURE 9.2. The point of the field. (Drawing by the author.)

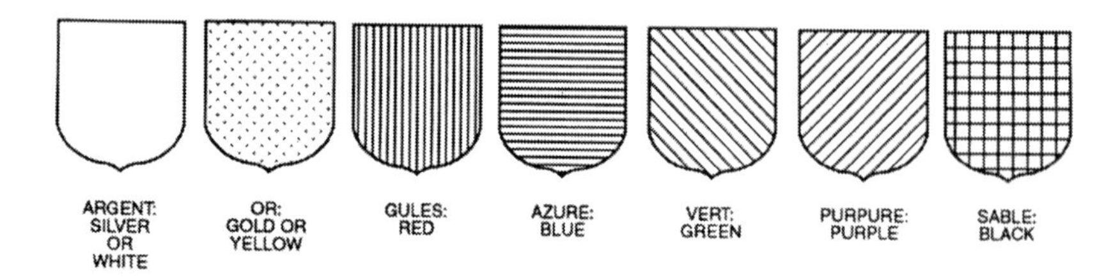

FIGURE 9.3. Tricks are black and white cross hatching (originally developed by engravers) to represent tinctures when colored pigments are not available. (Drawing by Matt Alexander and Rosmarie Awad.)

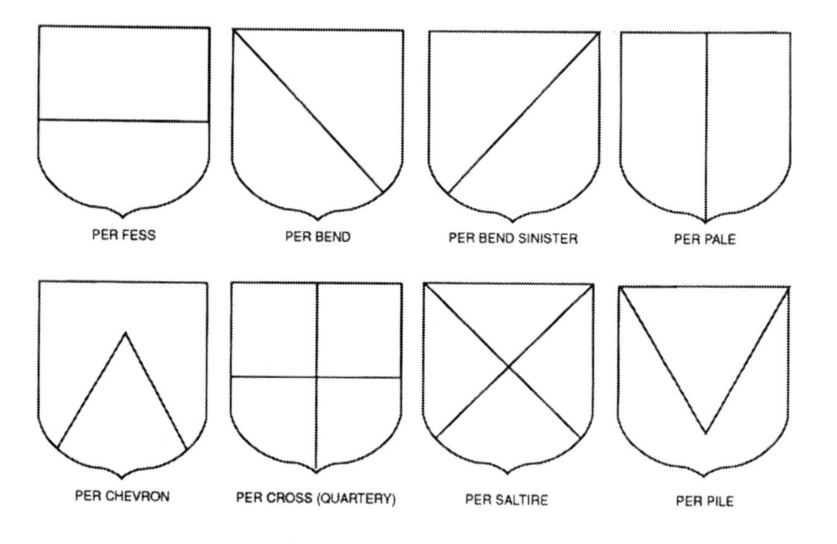

FIGURE 9.4. Ordinaries. The party function divides the shield. (Drawing by Matt Alexander and Rosmarie Awad.)

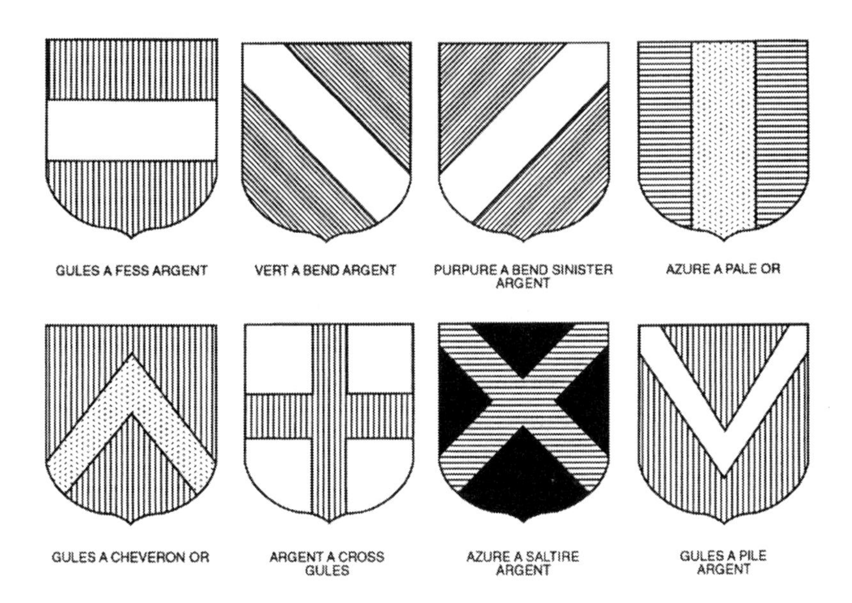

FIGURE 9.5. Third Ordinaries. (Drawing by Matt Alexander and Rosmarie Awad.)

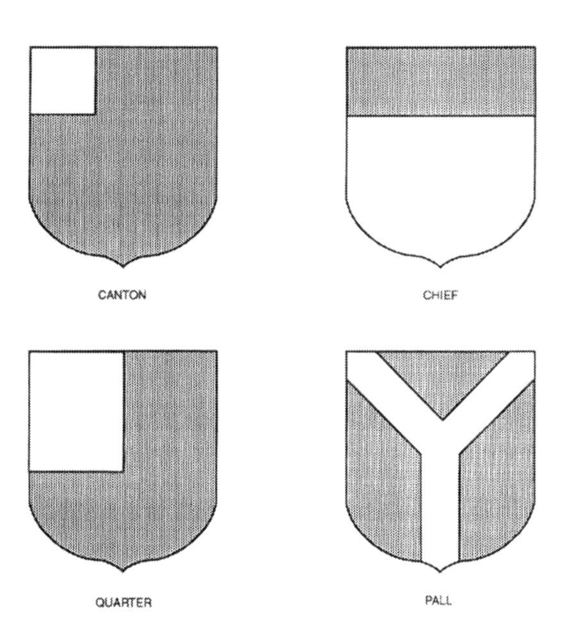

FIGURE 9.6. Sub-Ordinaries. (Drawing by Matt Alexander.)

FIGURE 9.7. Discrete charges. (Drawing by Matt Alexander and Rosmarie Awad.)

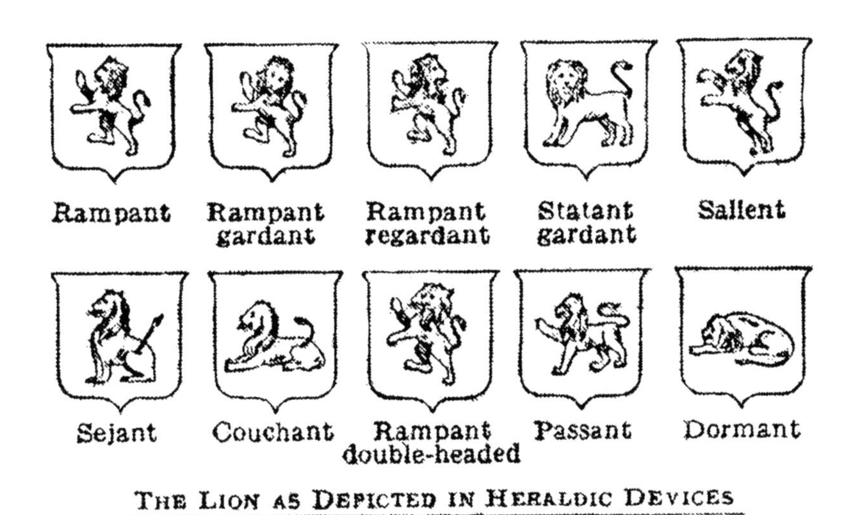

FIGURE 9.8. Lions. (Scanned from The Winston Simplified Dictionary, Philadelphia, 1929, and enhanced by Matt Alexander.)

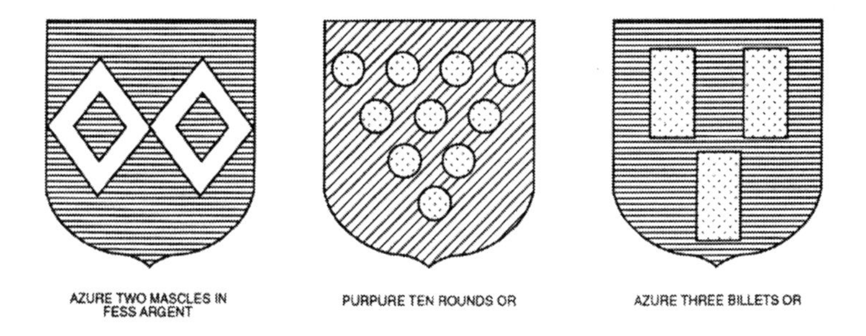

FIGURE 9.9. Discrete charges repeated. (Drawing by Matt Alexander and Rosmarie Awad.)

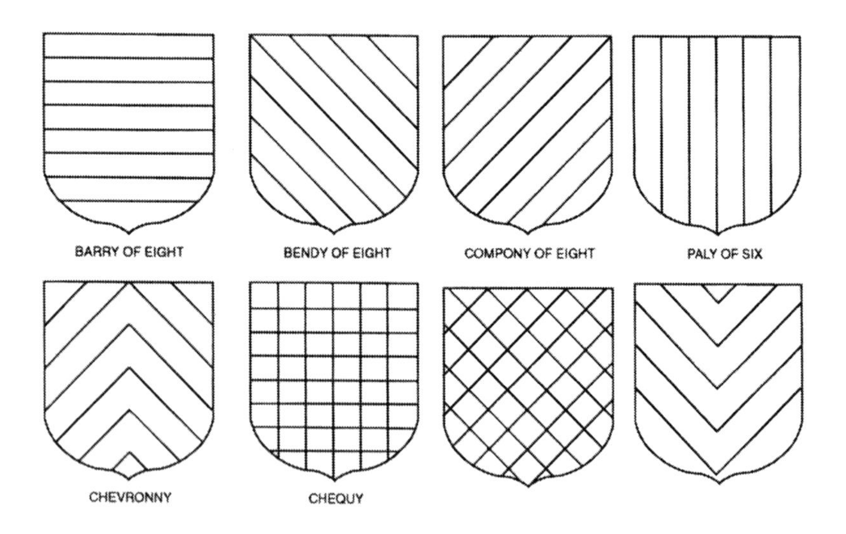

FIGURE 9.10. Ordinaries repeated. (Drawing by Matt Alexander.)

FIGURE 9.11. Many discrete charges. (Drawing by Matt Alexander and Rosmarie Awad.)

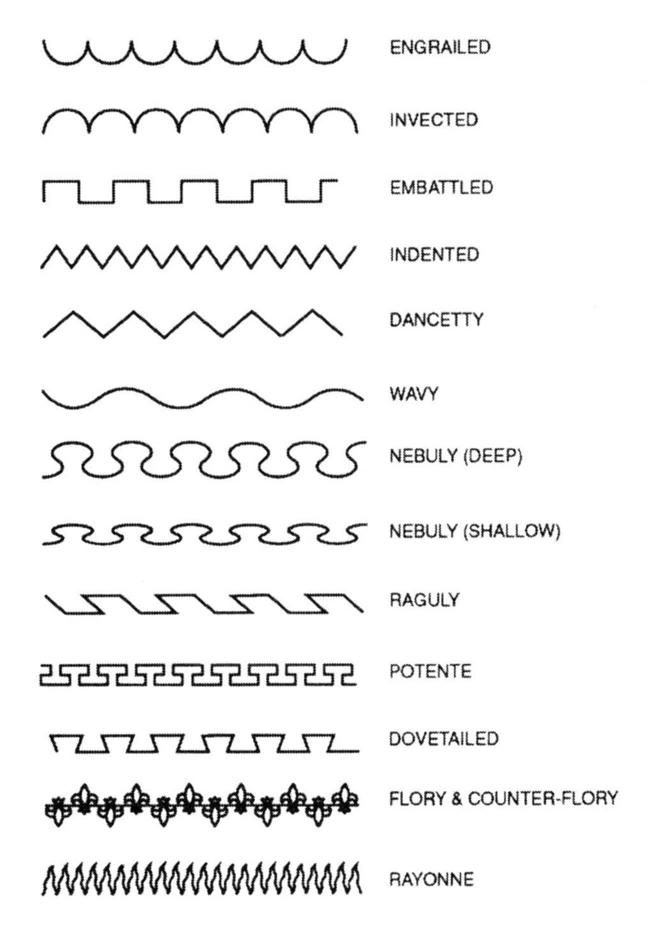

FIGURE 9.12. Partition lines. (Drawing by Matt Alexander.)

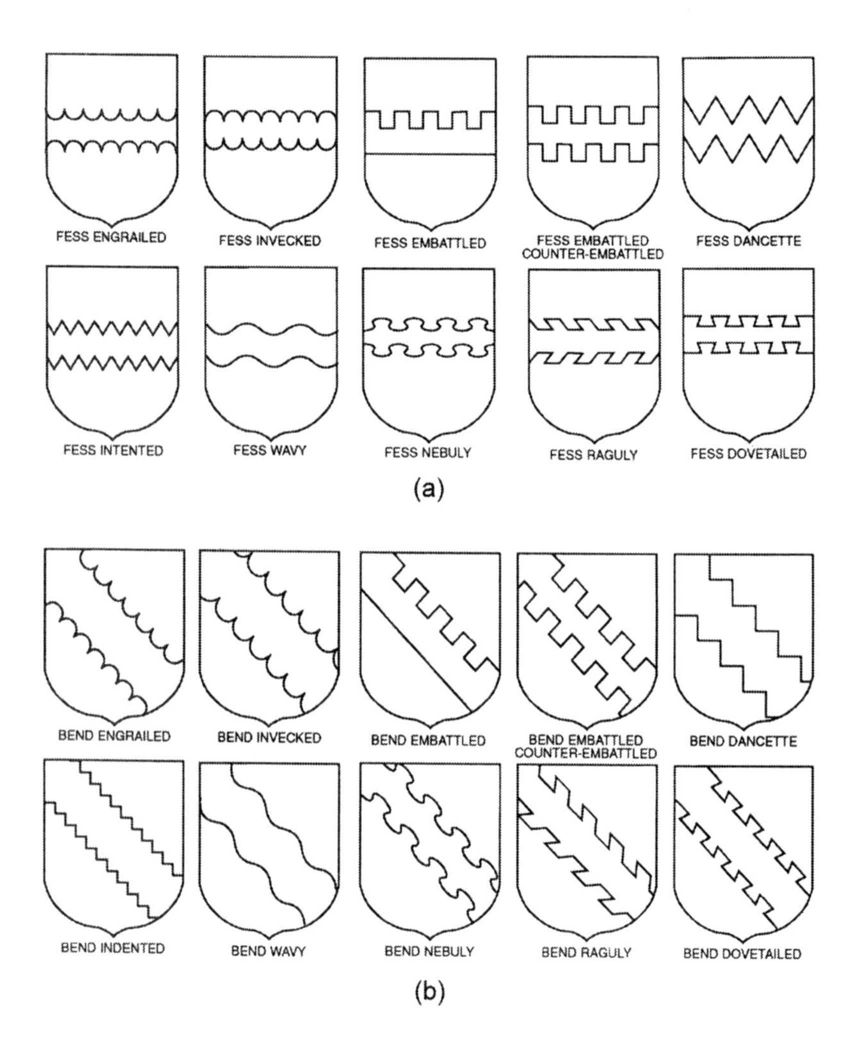

FIGURE 9.13. a) Fesses. b) Bend. (Drawing by Matt Alexander and Rosmarie Awad.)

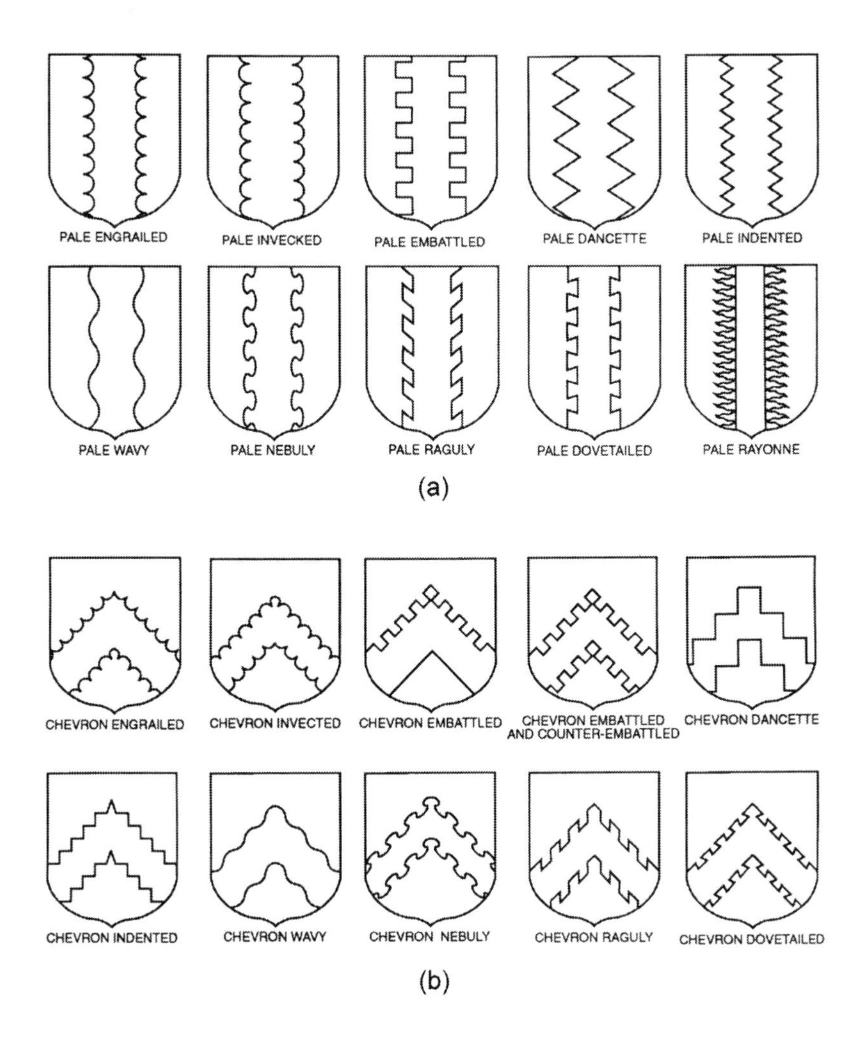

FIGURE 9.14. a) Pale. b) Chevron. (Drawing by Matt Alexander and Rosmarie Awad.)

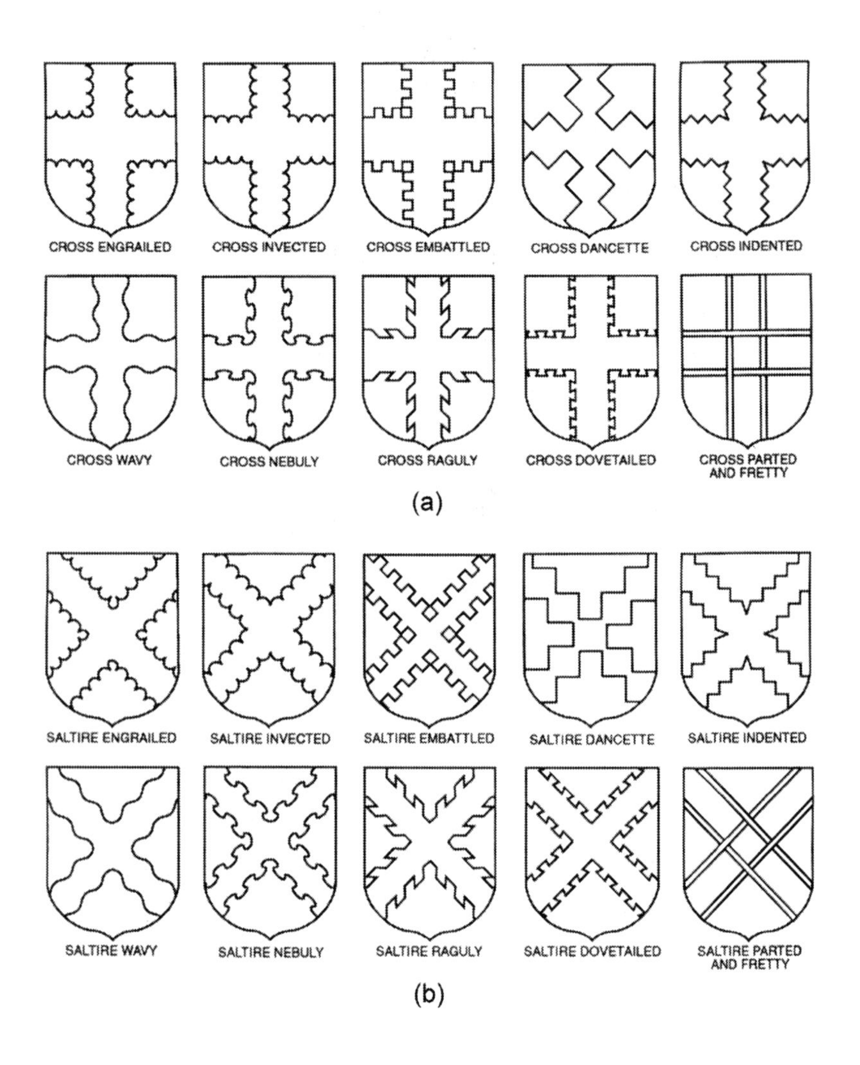

FIGURE 9.15. a) Cross. b) Saltaire. (Drawing by Matt Alexander and Marie Awad.)

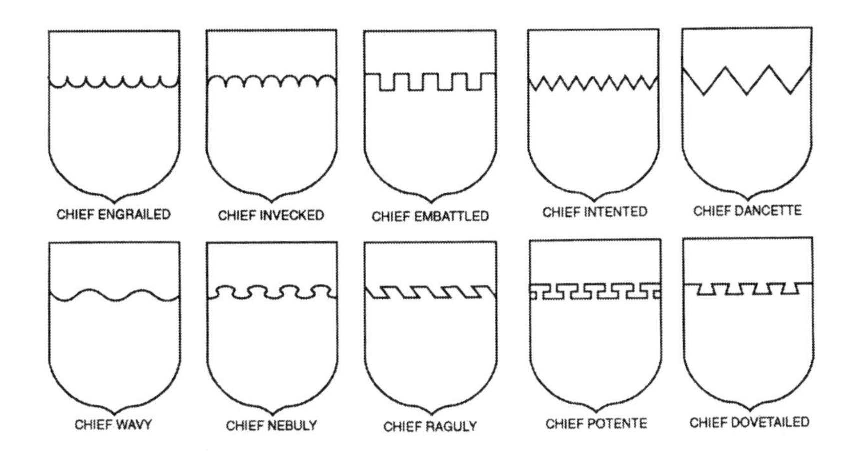

FIGURE 9.16. Chief. (Drawing by Matt Alexander.)

FIGURE 9.17. Marshalling of Arms. (Drawing by Matt Alexander and Rosmarie Awad.)

10

Masaccio's Bag of Tricks

Marc De Mey[1,2]

10.1 Introduction

Up to now, it has been fashionable, especially in artistic circles, to look upon linear perspective as a rather worn out, if not obsolete pictorial technique. Nineteenth century innovative painters had dismissed it as unessential. A pioneer like Cèzanne took it as a challenge to provide suggestive renderings of volumes while purposely violating the traditional rules of linear perspective. For the 20th century modernist, these nineteenth century inflictions upon the principles of perspective became a symbol of liberation, allowing for exploration and experimentation with new techniques and new themes in unprecedented ways.

The obsolescence of perspective in painting is in contrast to its restoration in recent theories of vision. In his influential stage model for an integrative theory of vision, David Marr [1] inserted a 2.5-D stage as his most original contribution. This 2.5-D stage is a pivotal step in assembling the perceptual object, being the initial global integration of the various computed features into a coherent whole. It is, however, a frozen view, indissolubly tied to the fixed standpoint of the viewer. It has depth already, but no volume yet, consisting only of oriented surfaces. Volume enters in stage three where coordinate system reference lines are located in the objects, coinciding with their main axes. Marr's 2.5-D only contains one coordinate system and its reference lines coincide with the main axes of the viewer's body. In a sense, it is also a restoration or a recognition of Vitruvius Pollio's reverence for the centrality of the body of man, not in its proportions but in its axes.

---

[1]University of Ghent,Biandijnberg 2, 9000 Gent, Belgium

[2]Versions of this paper have also been presented at the XIXth International Congress of the History of Science, Zaragoza, Spain, August 24, 1993 and at the Instituto e Museo di Storia Della Scienza in Florence, Italy, September 20,1993

The viewer's coordinates allow for the first overall integration and this viewpoint tied conception is considered an essential step on the way to complete spatial understanding. Thus, in a sense, Marr's major innovation is the rediscovery of perspective in a central role for perception. It is only a preliminary stage to full volume, but nevertheless essential scaffolding for erecting 3-D. Thus, the coherence obtained with the obsolete painters' procedure happens to correspond to the coherence exhibited in an essential component of every perceptual process.

Marr's most innovative stage is also his most controversial one. Critics and authors of alternative schemes such as Lowe [2][3] and Biederman [4] have argued that the 2.5-D stage is superfluous. Many perceptual modules would compute 3D interpretations straightforwardly from plain 2-D data, without any need for an intermediate integration of global spatial layout. The most liberated view is probably Ramachandran's [5], who considers vision a *bag of tricks*, a collection of selectively and variably recombinable perceptual modules. This "bag-of-tricks" view is the antipode of Marr's view, reducing perspective to one trick aside several other perceptual devices.

To what degree then is perspective this super-integrative mechanism that Marr saw in it? Is it the pivotal unifying scheme without which spatial interpretation of visual data is impossible? Or can it be reduced to one module of many others, easy to ignore or to overrule, as Cèzanne did, without basically endangering the integrity of the perceptual process? Is it reducible to one single trick in the impressive collection of tricks that seems to constitute our system for vision?

Let us explore the question by going back to a historical case in the discovery of linear perspective: Masaccio's fresco of 1425 depicting the *Trinity* which is to be found in the church of Santa Maria Novella in Florence. It is the oldest surviving painting considered to exhibit unambiguously an almost perfect adherence to the rules of linear perspective. It is the prototypical perspective painting [6], setting very high standards with a difficult rendering of a cylindric coffer vault. When Alberti wrote the very first treatise on perspective in his *Depictura* ten years later[7], he wisely recommended a flat square-tiled floor as a handy grid for the painter to subdivide and orient in the newly conquered space. In Masaccio's *Trinity*, being almost at eye level of the viewer, the chapel floor reduces to a horizontal line and the structure of space comes down from the more complicated but supposedly also square coffers of the curved ceiling.

We will first consider the painting and some models of its spatial layout. We will then focus upon the mechanisms of its magic and investigate whether perspective can account for the whole scope of powerful effects that the fresco makes. Thereupon, we can come back to the issue of perspective's pivotal role and its possible relationship with science and theories of vision in both Masaccio's days and our own time.

10.2 The Perspective and Plan of Masaccio's *Trinity*

Masaccio's *Trinity* depicts the central figures of the Christian faith in the rather common arrangements of the throne of Grace: the crucified Christ supported by God the Father, holding the cross with both arms, and in between their faces the shape of a dove representing the Holy Ghost. Standing beside the cross are the equally common figures of Mary, Christ's mother at his right and, at his left the most faithful disciple: Saint John. *Kneeling in front of the chapel, one at each side, are the two donorsin profile, a man and wife*(Figure 10.1).

Simson [8] has traced the iconographic theme of the *Throne of Grace* and provides a variety of examples which illustrate that the theme as such is not revolutionary new. Novel in Masaccio's version is the very realistic settings of the chapel and the real life size human bodies to depict the divine figures as well as the mortals.

The fresco is painted on the left wall of the aisle of the spacious gothic church of the Dominicans. The scene with the human figures is situated in a room that looks like a real chapel. Its architecture is closely related to the innovations Brunelleschi introduced in the building Spedale degli Innocenti. Its portico comprises a combination of similar arches supported by Ionic columns, flanked by pilasters with Corinthian capitals and medallions, all resembling the ones that Masaccio's depicted.

In his activities as an architect, Brunelleschi had almost completed this part of the building by the time Masaccio was about to start the *Trinity* fresco. Also in that same period, Donatello was giving shape to niches to accommodate some of his statues with the same architectural elements in very similar composition. Thus, there is

FIGURE 10.1.

no doubt that Masaccio's architectural components were real. They belonged to the innovative use of classical building components that Brunelleschi was introducing in Florence at that very time. And as indicated, the human figures are also very realistic in their carefully sculptured lifesize bodies.

Masaccio's *Trinity* has fascinated not only students of art history but also historians of science [9]. The newly discovered linear perspective seems so correct and so essential to the powerful impression it makes that scientists and mathematicians have felt tempted to study it as a genuine scientific discovery. Is it so meticulously correct?

10.2.1 The Ground Plan of G. F. Kern (1913)

The extensive literature on Masaccio's *Trinity* contains several contributions which propose plans for the imaginary chapel. We will only briefly touch upon the most important ones, if only to indicate how interpretations can still diverge in a case that seems at first sight almost overspecified in its measures.

After some penetrating studies of perspective in Northern painters like Van Eyck in 1904 [10], Kern [11] turned his attention toward the innovators of the South and provided a geometric analysis of Masaccio's fresco which included a ground plan of the chapel.

After a series of measurements and inferences, Kern concluded that the ground area covered by the barrel vault should be square. He felt supported by the sketches of a renaissance architect in a sixteenth century manuscript at the Ambrosiana in Milan. One of these sketches depicts a Roman sepulchral monument exhibiting some similarities with Masaccio's chapel. The ground plan is indeed square but the coffers of the barrel vault are considered to be rectangular.

With respect to Masaccio's chapel Kern struggles with an apparent dilemma of choosing between a square ground plan and square coffers for the vault since, according to his scheme, it is impossible to have both. He definitely chose the square ground plan for the global layout and proposed rectangular shapes of different sizes for the segmentation into coffers.

Kern relies on the abaci on top of the Ionic capitals to derive a viewer's distance. Despite the inaccuracy of the method he tries to determine the location where the elongations of the diagonals of the abaci meet with the horizon line. However, there is more uncertainty to Kern's data than the method could produce. As Janson [12] indicated, there are some puzzling inconsistencies in Kern's measurements which must derive from errors of transcription and which make some of his specific proposals very difficult to accept. However, Kern also makes some penetrating qualitative remarks with respect to the human figures in the fresco.

He notices that the rules of perspective are not applied with respect to the depicted persons. The figure of God the Father is too large given a position which Kern locates in the back of the chapel. Leaving aside this strange and remarkable positioning of the Father, Kern is right in noticing that the figures are not foreshortened. Despite their differing distances from the viewer, compared to the kneeling donors, Mary and John are similar in size. It is a rather daring step to conclude from this that the author of the architecture of the chapel has to be different from the author of the figures. Nevertheless, Kern considers the differences so pronounced that this conclusion cannot be avoided: he sees Masaccio as the author of the figures and assumes Brunelleschi to be the author of the chapel architecture.

10.2.2 The Ground Plan of Sanpaolesi (1962)

In a book on Brunelleschi[13], Sanpaolesi proposes a plan of Masaccio's *Trinity* chapel with a similar square ground plan. However, he attaches the support of God the Father to a sarcophagus-like structure so that it now reaches as far as the middle of the chapel. This avoids the awkward discrepancy which Kern thought to detect in the position of the Father's feet necessarily back in the chapel and his head and shoulders way up in front. The Father can now be located in an upright position under the fifth rib of the barrel vault, still in the back part of the chapel, but just behind the middle. In fact, Sanpaolesi places the cross right in the middle of the chapel.

It seems a very plausible model but it remains incompatible, like Kern's, with square coffers. Also, it places the viewer relatively close to the frontal arch of the chapel at a little more than five meters distance. Also, the large structure to which the Father's support is attached remains rather ambiguous. If it is to be seen as a sarcophagus, when parallel to the back wall, could it fit between the columns supporting the barrel vault? As Janson ([12], p 84) indicates: a sarcophagus between the columns supporting the arch is highly improbable since it would be too short to contain a regularly sized body. In his monograph on *The Psychology ofPerspective and Renaissance Art* [14], Kubovy chooses Sanpaolesi's elevation because it shows the location of the figures. Alternative plans with a somewhat deeper space, such as Janson's [12], provide room for both the figures and a sepulchral tomb of which the long axis is in line with the main axis of the chapel.

10.2.3 Janson's Ground Plan (1967)

As indicated by Field, Lunardi, and Settle [15], a method based upon the ribs of the barrel vault with coffers considered as having square shapes allows for the most plausible viewing distance. It is much more accurate than the method based upon the abac. It locates the viewer on the separation between the aisle and the nave, at a distance of 6.86 meters from the wall with the fresco, substantially further than in Sanpaolesi's scheme.

Janson manages to preserve a square shape for both the coffers and for the whole chapel. However, the square ground plan is no longer located exactly under the vault. The square is defined by the four columns supporting the two cornices which are situated parallel to the central axis of the chapel and which support the barrel vault. On the fresco itself, only few features provide indications for these elements. The frontal columns are hidden behind the pilasters and the ones in the back are barely visible, hidden by the frontal columns which support the arch. Nevertheless, these almost hidden columns reveal themselves through fragments of the volutes of their Ionic capitals. Also, the cornices which connect the frontal outward columns with the back ones along the sides of the chapel are partly discernable, rather conspicuously with their steep vanishing line slant. Together, the features suggest the additional space in width which Kern only accounted for in terms of niches. The gain in width allows Janson to increase depth without sacrificing the overall square shape.

It is not clear how serious this requirement for a square ground plan should be taken. Janson ([12], p 84) claims that "we must... postulate that the plan of the chapel was a square; a less regular form would be inconsistent with the solemnity of the subject." Were imaginary chapels subjected to much more strict geometric requirements than the ones that were really executed in brick? Both the famous ducal palaces of Mantua and Urbino contain house chapels which have a rectangular shape. But on the whole, Janson's square ground plan allows for a consistent reading of Masaccio's Trinity.

10.2.4 Kemp's Plans for the *Trinity* (1981, 1990)

In his 1981 book on Leonardo da Vinci [16], Kemp provides a plan of the *Trinity* chapel that resembles Janson's in general outline. It has almost the same square as basis though it differs in the detailed arrangement of the abacuses which Janson had cut off in a slightly different way. This plan shows the location of the figures and indicates the position of a tomb to which the footboard of God the Father is attached.

In view of the length of the footboard, the tomb would be rather short unless extended to the back of the chapel, a possibility the dashed lines could probably accommodate.

In his epoch-making book *The Science of Art* [17], Kemp surprisingly comes back to the original Kern [11] concept of a square basis coextensive with the area covered by the barrel vault. The elevation provided with the plan indicates the shorter viewing distance of 5.40 meters rather than the 6.86 meters suggested by Janson. This shortens the chapel accordingly and though the tomb is now clearly extended to the back of the chapel, it is doubtful whether in this room it can be made long enough to contain a normal sized body. Though Kemp indicates theapplication of some corrections on Janson's scheme either explicitly given or implied by Polzer's [18] detailed study of the surface of Masaccio's fresco, it is in no way obvious how they should lead to such an extensive modification.

The apparently unsettled issue of the ground plan for Masaccio's *Trinity* made us decide to develop a 3-D computer model of the chapel. Given the knowledge of the Brunelleschian architectural elements used in buildings such as the *Ospedale degli Innocenti* erected at the very same time, it should be possible to build an architectural model that would allow empirical exploration and verification of the various depths for the chapel and viewer distances that vary accordingly.

FIGURE 10.2.

10.2.5 Proposed Plan and Elevation for Masaccio's *Trinity*

A major measurement operation and critical analysis of proposed plans has been undertaken by Field, Lunardi, and Settle [15] who were allowed, like Polzer [18], a very close inspection of the fresco. They do not provide a detailed ground plan but, in general, they support a plan that closely resembles Janson's. They consider Kern's [11] abaci method inadequate and unreliable and base their own inferences and extrapolations upon the rib system of the barrel vault and the hypothesis of square coffers.

For our own first approach, we have taken our measures mainly from the rich collection of data in Field, Lunardi and Settle [15] augmented with those of Polzer [18] and some graphical suggestions of Schlegel [19]. Without providing justifications for each measures in detail, we provide line drawings for plan, elevation and frontal view derived directly from our computer reconstruction.

In principle, several data for the frontal view can be measured directly within the picture plane (Figure 10.2).

The frontal arch which is within the picture plane measures according to Field, Lunardi, and Settle, 211.55 centimeters. This comes down to a circumference for the vault of 332.13 centimeters. As the circumference of the cylindrical vault is segmented into eight coffers, divided by eight, this yields 41.51 centimeters per coffer (including ribs). Once decided upon square shaped coffers, this information is sufficient for a structuralization of the ground plan.

The vault is seven coffers deep. With sides of 41.5 centimeters this yields as total depth for the vaulted area 290.5 centimeters, again all ribs included. It is plausible to suppose that Masaccio did not choose the depth of the two archs to be different from the depth of the coffers so that the total depth from the front of the first arch to the back of the second arch reaches 373.5 centimeters. The measures are slightly larger than Polzer's and Field, Lunardi, and Settle's. This partly depends on decisions with respect to whether to accept full or half ribs at the locus of contact with the cornices. The measure of 41.5 centimeters slightly overextends the size of the abaci which Field, Lunardi, and Settle take to have sides of 41 centimeters. But on the whole, these easily derived measures fit rather well with the requirement of organization in the shape of squares for both the global ground plan and the structural parts (Figure 10.3).

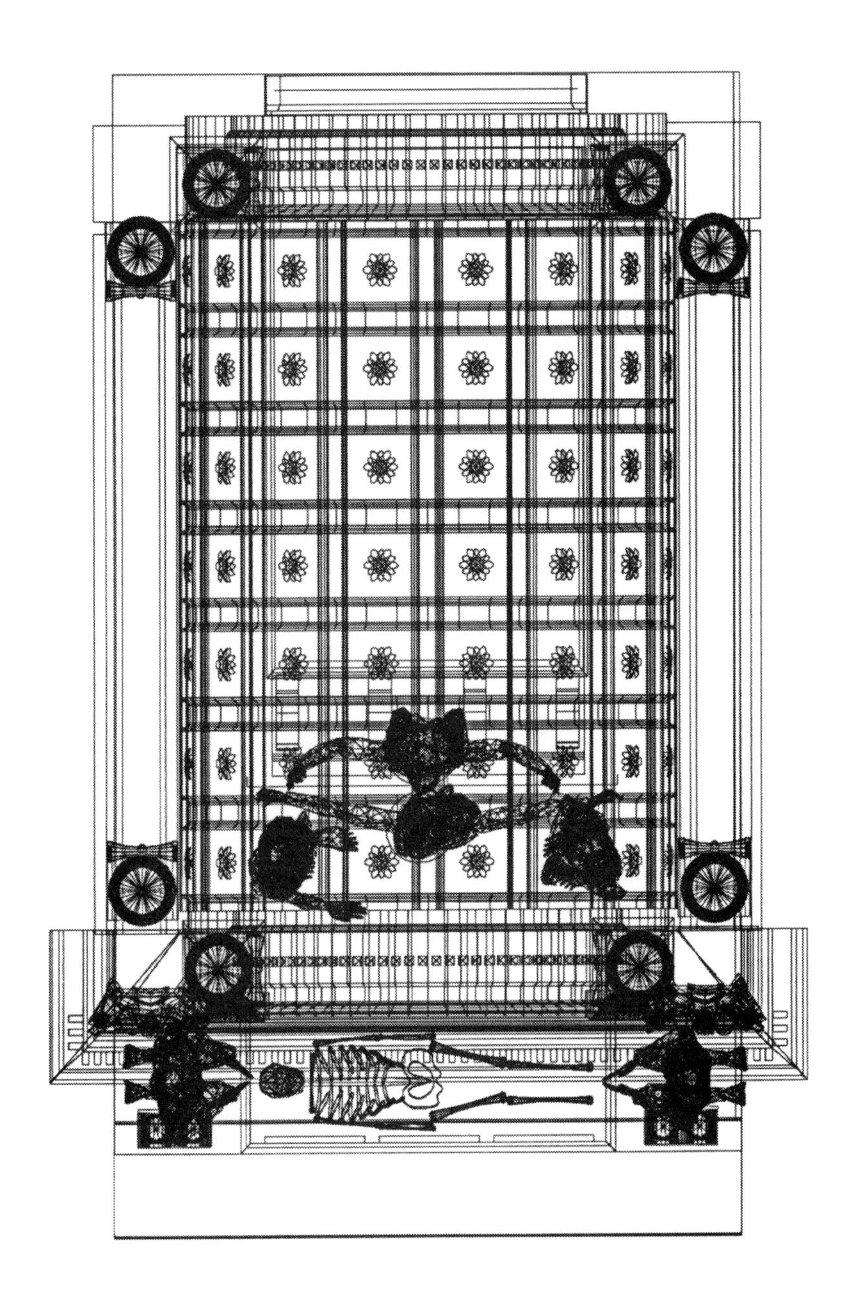

FIGURE 10.3.

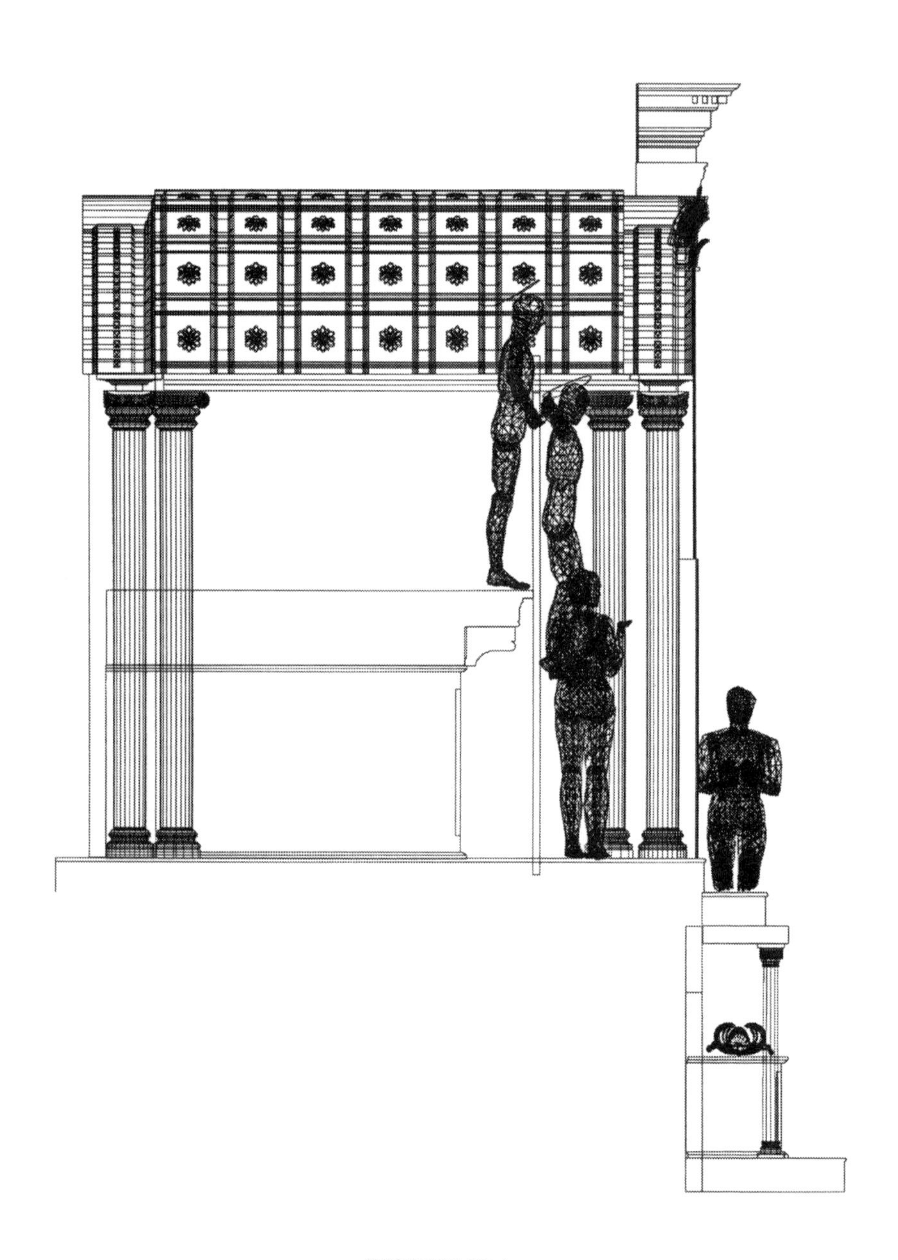

FIGURE 10.4.

By inspecting the elevation one can see that the chapel can now easily contain a tomb of 200 centimeters in length or longer. However, when looking at the figures, one realizes to what extent the location of the viewer can make a difference. Compared to Sanpaolesi's model, which takes a closer viewing distance, our chapel is drastically stretched backward. God the Father is in a frontal position, already under the second coffer. In Sanpaolesi's compressed chapel, the Father is located much further in the back, as far as the fifth semicircular rib (Figure 10.4).

That we can derive a perspective view from our computer model that comes quite close to the linear qualities of Masaccio's fresco is no proof for the correctness of our model. Based upon a shorter viewing distance, a shorter model of the chapel could provide almost identical views, though at the price of sacrificing the square shape of the coffers, and eventually of the ground plan. However, as the points we want to illustrate with the model relate mainly to Masaccio's rhetorical use of the perspective scheme, the ultimate correctness of the reconstruction is not a crucial issue here. As Masaccio's use of perspective goes beyond the straightforward application, any implementation of a computer model would ultimately fail as a complete reconstruction of his achievement.

Only to illustrate that, based upon a viewing distance of 6.86 meters, the computer model indeed yields the Masaccio type of view, we generate it in line drawing for a beholder of 172 centimeters tall, fixating the vanishing point at the same height from the floor right in the middle of the picture surface (Figure 10.5).

To indicate that any other standpoint is equally within reach, we also generate a view from an entirely different location: the position looking downwards upon the cross which Saint John of the Cross used to depict a dream. It is the rather strange angle upon the crucified Christ which inspired Dali in his Christ of Saint John of the Cross (Figure 10.6).

10.3 Convincing Realism or Subtle Rhetoric Tricks?

The architecture and arrangement of figures in the *Trinity* have an unmistakable grip upon the beholder. The scene spreads abroad an atmosphere of solemnity and serenity. Irresistibly, one is drawn into

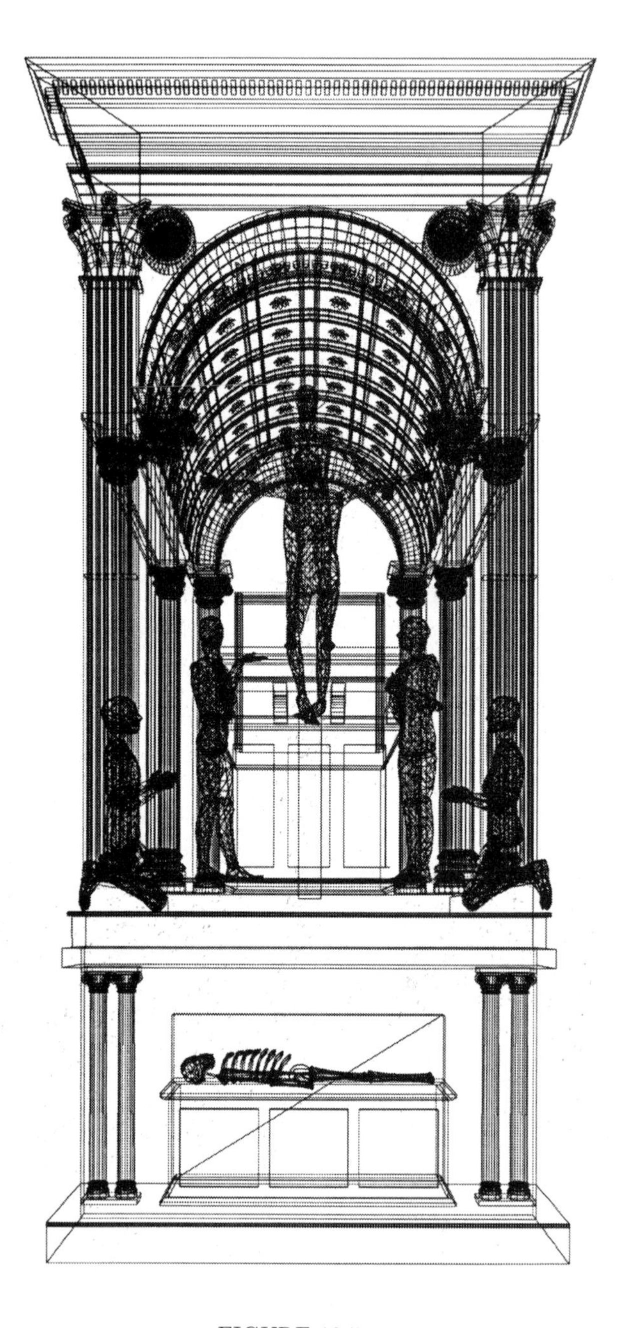

FIGURE 10.5.

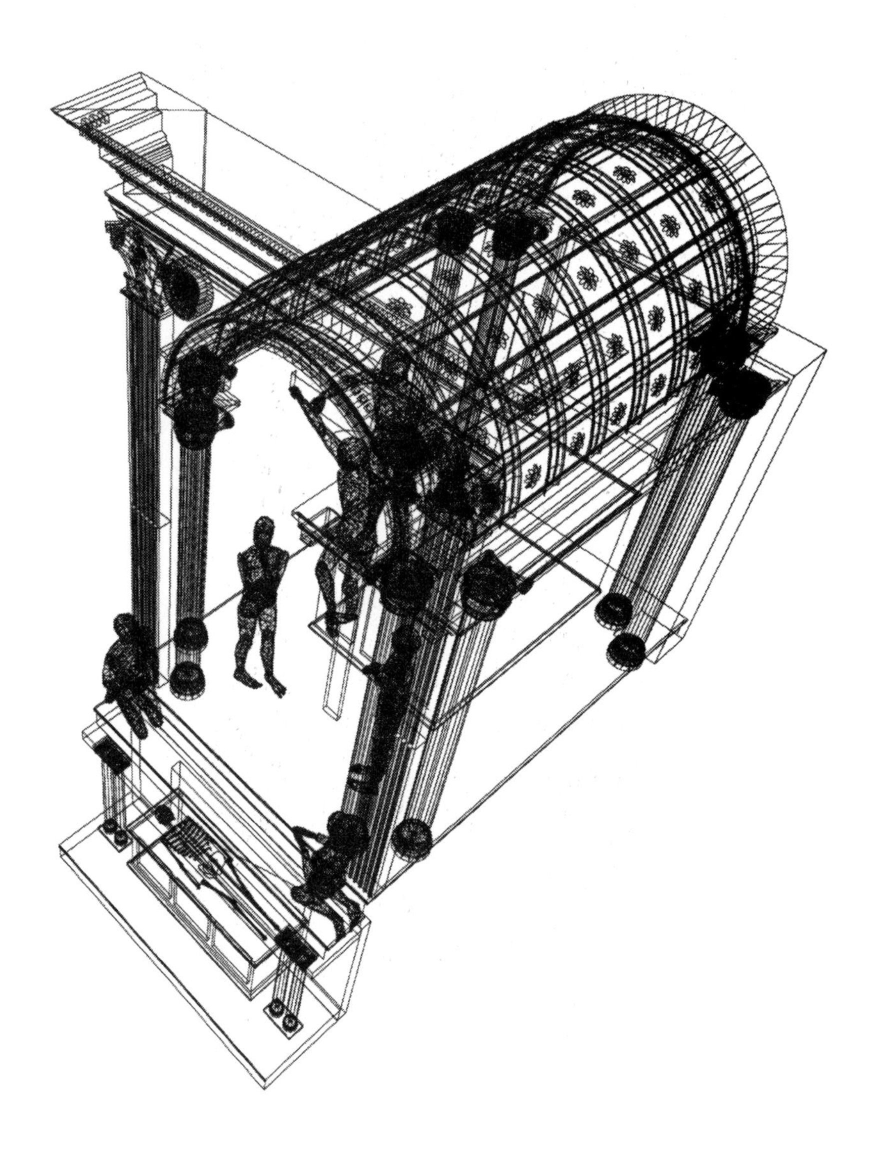

FIGURE 10.6.

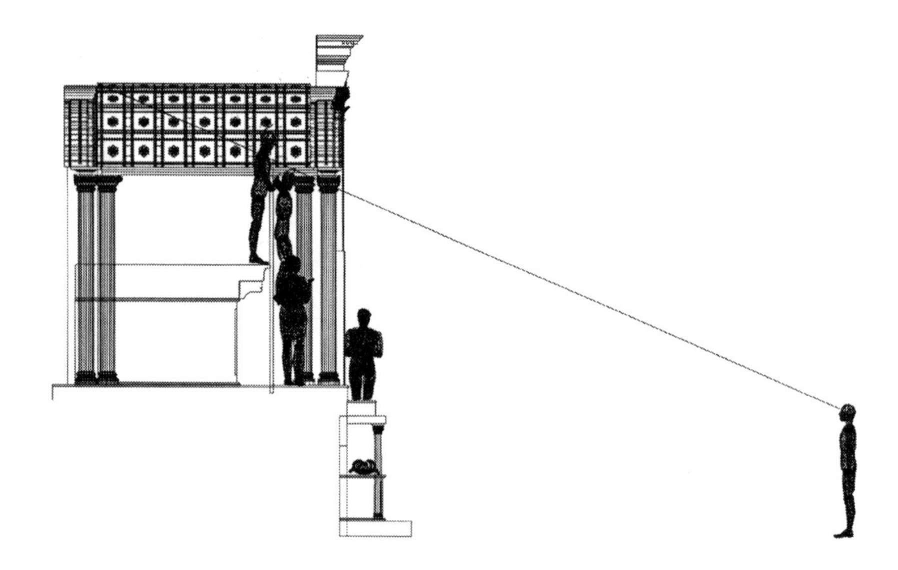

FIGURE 10.7.

eye contact with the Father who looks down upon the viewer from his high position with slightly sad and forgiving eyes. The absence of any fierceness or blame in these eyes seems in sharp contrast with the overwhelming power of the scene. Is its strength due to the control of space with the choice of a viewpoint that amplifies the monumental character of the arch and expands the spaciousness of the vault?

10.3.1 Control of Space

Whether looked upon from the contracted Sanpaolesi elevation or from our own elongated chapel reconstruction, the head of God the Father is almost on the diagonal that connects the middle point of the frontal semicircle of the vault cylinder to the top location on the circumference of its base in the back. This is a line of sight which follows a path that is among the longest possible trajectories through the vaulting space. If it is the space behind God the Father in particular that contributes to the impressiveness of the scene, the effect is stronger to the degree that the Father is in a more frontal position. In this sense, the spaciousness is dramatically greater for the larger model of the chapel (Figure 10.7).

Although we can argue that the effect is due to a perfect control of space, it cannot be entirely reduced to it. Obviously, the masterly control of the space of the vault is of crucial importance for this effect to work. However, the majestic magnanimousness also results from the contrast between the modest size of the Father's head and the large space that surrounds it. The complementarity between the small space of the Father's head and the extensive external space is nicely indicated by the curvature of the arch which resonates the curvature of the Father's head and the curvature of his nimbus (Figure 10.8).

The concentric arrangement of the curves of the Father's head, the nimbus and the arch is an aspect of surface organization. In art historical and art critical literature, surface aspects relate to purely 2-D organizational patterns. They are entirely within Marr's first stage of two dimensional structuralization of the raw material of perception. Assignation of lines and detection of parallelism between lines or concentricity between curves belong to this primary level of processing. The role of *surfaceorganization* in this case already hints at a peculiar possibility for artists to amplify advanced 2.5-D and 3-D effects by means of elementary 2-D mechanisms. Before exploring this possibility any further, it is useful to notice a few cases where Masaccio even appears to overrule 2.5-D requirements for favored 2-D patterns.

10.3.2 Concessions to Surface Arrangement

Whereas Masaccio's *Trinity* used to be referred to for a long time as an impeccable and prototypical application of the rules of perspective, more recent art historical literature has increasingly focused upon some inflictions of Masaccio upon these rules. Both Field, Lunardi, and Settle [15] and Kemp [17] point out a number of aberrations not all of which could be due to ignorance or negligence by the painter.

The fact that we are confronted with two rivalling systems for deriving the perspective structure already indicates a lack of coherence. The method of the abaci yields a closer viewing distance than the method of the ribs and the coffers. The dominating trend was to distrust the abaci as reliable sources for the spatial information. They were considered as added-on structures which had no pivotal role in the primary construction.

Other apparent aberrations are less clearly the result of inaccu-

FIGURE 10.8.

racy or unimportance. Kemp ([17], p 20) includes the slightly tilted position of God the Father to his list of inconsistencies: "The apparently centralised symmetry of God the Father and the crucified Christ is subverted by marginal shifts away from the axis, as in God's head, which is just to the left of centre." Field, Lunardi, and Settle ([15], p 35) too have noticed this unbalance: God the Father is 2.6 centimeters out of the center of the picture. Instead of seeing this as an unintended side-effect of inaccurate methods, one can also consider this as a subtle means to evoke more tension. As Field et al. indicate, the aberration is too pronounced to be considered as unintended. Why should it not be interpreted as the expression of dynamic tension by means of tipping volume axes, a technique masterly exploited by Cézanne as convincingly demonstrated by Loran [20]? In his discussion of Cézanne's *Portrait of Madame Cézanne,* Loran illustrates how the volume axis of the main figure is falling to the left.

In addition to the dynamics of surface and space emphasized by Loran, such subtle imbalances are also evocative of mass and volume. In the case of God the Father it is evocative of both the physical force involved in supporting the cross and the gravitational pull on a body that is slightly out of balance. Notice that the linear elements involved here are the central axes of the volumetric units which typify Marr's full 3-D stage.

A deliberate deviation from expected regularity are the differing sizes of the coffers which stand on the cornices at both sides of the vault. The six coffers in the middle each subtend about 20 degrees of an arc, while the lateral ones subtend about 30 degrees. Why these elongated forms at the sides? Was Masaccio anxious to preserve some of the squareness of the coffers which became apparently endangered by foreshortening? By reducing the width of the most visible coffers, they looked less rectangular and the most frontal ones in the middle definitely preserved the characteristics of a square. On the computer reconstruction in which we have preserved a regular segmentation of eight coffers, each subtending 22.5 degrees to fill the 180 degrees of the arch, all the coffers look rectangular including the central ones in the frontal area of the vault.

Another interpretation, not necessarily in conflict with the previous one, is that the lateral coffers were changed for allowing the line of their transversal ribs to join the vanishing lines of the volutes.

In this way, these volutes and the farthest parts of the arms of the cross and Christ's hands come to lie on two major vanishing lines. This is again a somewhat weird surface adaptation: local correctness from a 3-D point of view is sacrificed to permit a 2-D-type linear organization which ultimately has a strong 2.5-D significance.

Other subtle surface patterns work out remarkably well seemingly without interfering with the spatial organization of depth. This is in particular true for the central position of Christ whose head is in the middle of the whole area covered by the chapel and in the middle of the trapezium defined by the four visible Ionic capitals.

It should be clear that the rules of perspective do not constitute the sole set of principles which governs Masaccio's construction of the chapel and its figures as a perceptual object. Marr's three types of processes: surface organization (2-D), perspective (2.5-D) and volume orientation (3-D) all seem involved, almost on an equal base, in providing a pictorial rendering of a perceptual scene. In Marr's scheme, they constitute a logical sequence. Is there no logic in their pictorial application?

10.3.3 Amplification of the Newly Discovered Device

When analyzing pictures by means of Marr's categories, we can distinguish between three types of lines according to his three stages.

- Lines for surface organizational patterns entirely within the picture plane corresponding to the 2-D stage. We can include here the virtual lines of the Kanizsa-type which evolve out of local interaction between perceptual elements just like in Glass-patterns.

- Lines within the picture plane which indicate surfaces at an angle and, in early perspective paintings, preferably perpendicular to the picture plane. They are not the needles of Marr's needle image but they coincide with the slanted lines which extend easily into vanishing lines.

- Lines which indicate the central axis of volumes and which inform on the orientation of volumes within the pictorial space. The imaginary line indicating the tilted position of God the Father would fit this category.

As Field, Lunardi and Settle have indicated, if a major discovery is involved in Masaccio's developing of the perspective of the *Trinity*, it certainly is the *centric point* with the *vanishing lines*. The basic insight undoubtedly has been the notion that orientations perpendicular to the picture plane have to be represented by *linesconverging upon a single point*. Obviously, this corresponds to the discovery of the 2.5-D and the means to locate the position of the viewer in the picture. But equally noticed by Field et al. is that the *centricpoint in Masaccio's Trinity* **not only** plays a role in its **perspective** scheme but also in the organization of its **surface geometry**:

- We have already pointed out how the hands of Christ come to lie upon the vanishing lines of the volutes;
- The lines corresponding to the slope of Mary's and John's clothing too converge upon the vanishing point as noticed by Field et al.;
- Also the edges of the abaci markedly exhibit the converging pattern.

Although the convergence might be somewhat less exact than it seems (there is no single vanishing point but a small area in which vanishing lines meet), it is obviously a major organizational principle in the linear arrangement of the painting. Most remarkable however is a kind of **secondary amplification** of **vanishing lines** by means of virtual lines of a purely 2-D status. This is indeed the extension of the 2.5-D -type vanishing lines in 2-D-type surface organization lines.

Look at the remarkable way in which the virtual lines between the Ionic capitals of the frontal columns and the Corinthian capitals of the pilasters extend the genuine vanishing lines between the Ionic capitals in the back and those in front. By putting all the *ornaments on a straight line* Masaccio creates a *virtual line* which emphatically *amplifies* the *vanishing line* (Figure 10.9).

It is a tricky extension of a 2.5-D line with 2-D means. But Masaccio's move is even more intricate. The ornaments themselves exert a perspective effect through their diminishing sizes. Again, this is entirely appropriate according to the rules for the spatial relationship between the Ionic capitals. However, it is entirely coincidental between the Corinthian capitals and the Ionic ones in front since they almost share the picture plane. Seemingly by a happy accident the larger size of the Corinthian capital comes to dramatize the recession in depth of this ornament line. Again, an extra amplification of the vanishing line and its spatial significance.

This creative exploitation of ambiguity is not a personal trick of Masaccio. As White [21] has so convincingly illustrated, Giotto already loved to confuse the viewer with double status lines. In *The Wedding Feast atCana* (Arena chapel, Padua), he has one horizontal line represent with its middle segment an horizontal line on a plane parallel to the picture plane while the segments at the extremes represent the horizontal line on planes orthogonal to the picture plane.

Masaccio's contemporary Donatello too obviously cherishes this same ambiguity which is so manifest in his big medallion, *The Resurrectionof Drusiana.*

Masaccio's three ornaments on a straight line embody this same principle of this single line the segments of which have different functions and are even perpendicular to each other. As the centric point is apparently the major discovery during this period, Masaccio uses this line complication device to increase the intricacy of the linear elements he considers most important: the vanishing lines.

10.4 Conclusion

Massacio's amplification of the vanishing lines by means of the virtual lines between the Ionic and Corinthian capitals is a clever application of 2-D surface organizational means to amplify the 2.5-D effect of perspective. As a pictorial technique, it is less revolutionary than we might think since artists clearly love to cultivate ambiguity such that single elements fulfill multiple functions. The ambiguity of the single straight line as one of the simplest 2-D patterns which can fulfill intricate 2.5-D and 3-D functions is a most illustrative example. However, is this only a superimposed effect upon the application of linear perspective which could be considered a genuine revolution in its own right?

In a recent paper Elkins [22] disputes the notion of a unitary and coherent notion of Renaissance perspective. He convincingly illustrates the heterogeneity of methods and meanings hidden under that label. In his interpretation, the theory of linear perspective is an invention of mathematicians, retroactively read into the heterogeneous

FIGURE 10.9.

set of discoveries of Renaissance artists. Our exploration of Masaccio's innovation certainly confirms this impression of heterogeneity.

We have only touched upon a few aspects of Masaccio's perspective. We have pointed out his fascination with vanishing lines but we have ignored his restraint in the application of foreshortening. Possibly, he consciously kept the figures of the *Trinity* more comparable in size than the geometric laws of perspective and foreshortening impose. But a colleague like Mantegna, who a few decades later shared Masaccio's sensitivity for the power of amplified vanishing lines, was much more tempted to explore the effects of the latter technique. A famous Mantegna fresco with a dramatic perspective on a Roman triumphal arch is Saint James led to execution, until its destruction during the second World War to be found in the Emeritani church in Padua. We can illustrate the difference it makes by having our computer model of Masaccio's *Trinity* generate a view from a Mantegna-type of position. Such an exercise demonstrates that it is not perspective as such, in a straightforward application, but its integration in a unique combination of several artistic devices that is behind the specific impact of a given painting (Figure 10.10).

While all views are obviously within the mathematical possibilities, it should be clear that Masaccio distills from them according to a recipe that differs from Mantegna's. Foreshortening is part of Mantegna's recipe in a way it is strange to Masaccio's.

Similar arguments can be made for other Renaissance painters of the innovative perspective period such as Uccello or Piero della Francesca. Each of them embodies a personal interpretation of perspective. The impact of their art is not so impressive because of their convulsive adherence to one single set of tight rules but because of their capacity to play with ambiguity: to have a single pictorial element eloquently serve more than one function, whether spatial or spiritual or whatever. Artistic impact depends upon skillful preservation of ambiguity and multiplicity of viewpoints at various levels, even with apparently simple and straightforward presentations.

We did not find the one pivotal discovery which should be thought of as responsible for bringing unprecedented unity to pictorial space in Masaccio. His perspective is an efficient device which he handles with great power. But the fact that he amplifies and embellishes it with additional surface features shows its ultimate rhetorical nature. It is not the essential rational procedure which, once discovered, had

FIGURE 10.10.

to be applied in all pictorial representation. It is but one version of a set of possibilities for evoking spatial effects, one trick among a diversity of other tricks.

We went back to the Renaissance introduction of perspective in order to understand its fascination in some current theories of perception. The all encompassing pivotal and unitary scheme could not be found in the methods of art. Nevertheless, unity and coherence are not necessarily retroactively projected into it in the way Elkins suggests. More than a century and half before Masaccio an influential theory of visual perception became popular: the perspectivist doctrine based upon Al Haytham's *Optics*. According to Bergdolt [23] attempts to apply it in visual arts and painting already began shortly after major formulations by Peckham and Witelo around 1280. Possibly, the history of painting from Giotto onwards is also partly the history of a series of attempts to faithfully apply this powerful unitary doctrine. There might be a common source behind the rich diversity encountered. A unitary theory can be productive by eliciting or provoking a plurality of attempts for its application. In that sense, the late medieval perspectivist theory of vision could have been as productive in art, while basically wrong, as Marr's inspiring integration in science. Even if the final outcome turns out to be only a bag of tricks, it still is an impressive yield.

10.5 Acknowledgements

The author gratefully acknowledges the substantial help from Johan Donné for detailed analysis of the measures of Masaccio's *Trinity* and from Wim De Boever for its computer reconstruction and visualization.

10.6 References

[1] Marr, D. (1982). *Vision,* San Francisco: Freeman.

[2] Lowe, D.G. (1985). *Perceptual Organization and Visual Recognition.* Boston: Kluwer.

[3] Lowe, D.G. (1987). Three-dimensional object recognition from single two-dimensional images. *Artificial Intelligence*, 31, 355-395.

[4] Biederman, I. (1987). Recognition-by-components: A theory of human image understanding. *Psychological Review*, 94, 115-147.

[5] Ramachandran, V.S. (1990). Visual perception in people and machines; In: Blake, A. and Troscianko, T. (Eds.), *AI and the Eye.* Chichester: Wiley, 21-77.

[6] Kuhn, J.R. (1990). Measured appearances: documentation and design in early perspective drawing. *Journal of the Warburg and Courtauld Institutes*, 53, 114-132.

[7] Alberti, L.B. (1435). *On Painting.* (translated by C. Grayson, with introduction and notes by M. Kemp), London: Penguin, 1991.

[8] Simson, O. von (1962). Über die Bedeutung von Masaccios Trinitätsfresko in S. Maria Novella. *Jahrbuch der Berliner Museen*, 8, 119-159.

[9] Santillana, G. de (1973). Art et science dans la renaissance. In: Buck, A., Costabel, P. et al. (Eds.) *Sciences de la Renaissance.* Paris: Vrin.

[10] Kern, J.G. (1904). *Die Grundzüge der Linear-Perspektivistischen Darstellung in der Kunst der Gebrüder van Eyck und ihrer Schule.* Leipzig: Von Seemann.

[11] Kern, J.G.(1913). Das Dreifaltigkeitsfresko von S. Maria Novella, eine perspektivisch-arhitekturgeschichtliche Studie. *Jahrbuchder königlich preusssischen Kunstsammlungen*, 24, 36-58.

[12] Janson, H.W. (1967). Ground plan and elevation in Masaccio's *Trinity* fresco. In: Fraser, D., Hibbard, H. and Lewine, J. (Eds.) *Essays in the History of Art presented to Rudolf Wittkower.* London: Phaidon Press.

[13] Sanpaolesi, P. (1962). *Brunelleschi.* Milano,

[14] Kubovy, M. (1986). *The Psychology of Perspective andRenaissance Art.* Cambridge: Cambridge University Press.

[15] Field, J.V., Lunardi, R. and Settle, T.B. (1989). The perspective scheme of Masaccio's *Trinity* fresco. *Nuncius*, 31-118.

[16] Kemp, M. (1981). *Leonardo da Vinci: The MarvellousWorks of Nature and Man.* London: Dent and Sons.

[17] Kemp, M. (1990). *The Science of Art: Optical themes inWestern Art from Brunelleschi to Seurat.* New Haven: Yale University Press.

[18] Polzer, J. (1971). The anatomy of Masaccio's Holy *Trinity. Jahrbuch der Berliner Museen*, 13, 18-59.

[19] Schlegel, U. (1964). Observations on Masaccio's *Trinity* fresco in Santa Maria Novella. *The Art Bulletin*, 45, 19-33.

[20] Loran, E. (1943). *Cézanne's Composition: Analysis of His Form with Diagrams and Photographs of His Motifs.* Berkeley: University of California Press (1985 paperback edition).

[21] White, J. (1957). *The Birth and Rebirth of PictorialSpace.* London: Faber and Faber (1987 paperback edition).

[22] Elkins J. (1992). Renaissance perspectives. *Journal ofthe History of Ideas*, 53, 209-230.

[23] Bergdolt, K. (1989). Bacon und Giotto. Zum Einfluss der Franziskanischen Naturphilosophie auf die Bildende Kunst am Ende des 13. Jahrhunderts. *Medizinhistorisches Journal*, 24, 25-41.

11

Is Alligator Skin More Wrinkled Than Tree Bark?

The Role of Texture in Object Description

Nalini Bhushan[1]

A. Ravishankar Rao[2]

11.1 Introduction

Philosophers in the western world from the time of Democritus and Leucippus in the 5th century B.C. have been interested in the kinds of properties we attribute to physical objects. According to Democritus, the universe was composed of atoms and the void. Atoms had quite specific intrinsic properties - they were hooked and came in various sizes and shapes. They had solidity, and therefore had weight and position. They were indivisible, ungenerated, indestructible, and thereby changeless. Atoms constituted the matter of the universe and the void was what they moved through. As atoms fell randomly through the void they collided with other atoms, and the resulting entanglement led to the formation of macroscopic objects. In sharp contrast, properties such as being sweet, bitter, or colored, existed only "by courtesy" or by convention. These properties, unlike shape, size, weight, and position, were not part of the fabric of the universe, and so were not worthy of the same degree of interest as were the more fundamental, "real" properties.

In delineating properties in this way Democritus was in effect anticipating a distinction that would eventually engage the minds of scientists and philosophers for centuries. It was Robert Boyle in the

[1]Smith College, Department of Philosophy, Northampton, MA 01063.
[2]IBM, T.J.Watson Research Center, Yorktown Heights, NY 10598.

FIGURE 11.1. A selection of a few textures from Brodatz's album [8]. Identified top to bottom, left to right, they are: reptile skin, tree stump, grass lawn, homespun cloth, beach pebbles, brick wall, water, lace, European marble.

17th century who first gave the distinction the name by which it has now come to be known: *primary* and *secondary* [1].[3] But it was the British philosopher John Locke [2], in the second half of the seventeenth century, who made it famous. We have therefore chosen to follow Locke in our use of this distinction. Examples of Lockean primary properties are shape, size, solidity, and density; that of secondary properties are color, taste, smell, feel, and sound. The status of texture as a property, however, is not clear in Locke. Commentators, in so far as they have mentioned Locke's view of texture at all,

[3]It should be pointed out that Galileo and Descartes had also discussed the nature of 'qualitative' (i.e., secondary) properties.

have sometimes taken Locke to consider it as primary [3]; others [4] more recently as allocating to texture a special place, in that it is not primary in the straightforward sense, it has more of the attributes of primariness than secondariness. Our paper explores the distinction between primary and secondary properties as a way of getting a sharper understanding of texture.

Texture is an important visual cue that can be used to identify and distinguish different kinds of objects. Indeed, if we look around us, we find a world rich in texture: a typical indoor scene consists of patterned wallpaper, furniture with directional wood grain, and carpets with a regular weave, whereas an outdoor scene consists of the texture of clouds, patterns in the sand, and the waviness of grass. Figure 11.1 illustrates a few textures found in the world around us.

Following Locke's lead, however, the property of texture has not engaged the interest of philosophers. This is in sharp contrast to the property of color, for example [5, 6, 7, 8, 9]. The first reason for this is that philosophers were (and still are) interested in "simple" rather than "complex" properties of objects. Shape, size, and even color (which has been regarded as relational and therefore secondary), are all taken to be simples. Salient characteristics of simple properties are that they are homogeneous and are not reducible to any other property. Also, these are "local" properties of the object. Thus a point (or at least a pixel) can be both shaped and colored.[4] Texture, on the other hand, is a "non-local" property of an object. It is spread over a surface and composed of a combination of the simples such as size, shape, and density of the elements that constitute the object. This makes it a complex property.

Philosophers have been biased against complex properties (a) in their overriding concern with simple properties as revealing the fundamental nature of the object; and (b) in their reductive style of reasoning: if texture is nothing but simple properties in combination, then one can reduce away texture to its basic elements, and study those, the fundamental building blocks of objects, in order to get a fix on the nature of an object. If one wants to understand the

[4]Daniel Dennett makes use of recent empirical research in color vision to argue contrary to the intuitions of many philosophers and researchers in color vision that color is not a simple property, as Locke had assumed. That is, color is not a homogeneous power or disposition of a specific arrangement of microscopic textural properties of surfaces of objects. For Dennett, color is a relational but heterogeneous or complex dispositional property of objects [10].

nature of objects, one can do so without understanding the property of texture. Thus, the concern with simples, together with the belief in reduction, meant that texture was not of interest to philosophers. Our stance is in marked contrast to this. The claim that we intend to make good on is that texture stands as a high-level complex property in its own right. It is crucial both to our understanding of the fundamental nature of objects and to the way in which human beings actually go about identifying objects.

Researchers in certain areas of science, in contrast to philosophers, from psychologists and biologists working on perception and perceptual cues to researchers in artificial intelligence (e.g. [11, 12]), have historically taken texture to be both an important property of objects as well as a significant visual cue in perception and representation. One reason is that if it *is* a complex, non-local property of an object, in the sense articulated above, then identifying an object's texture takes one a long way toward identifying or recognizing the object itself. Despite this laudable history of research interest, it is fair to say that much of this work has been on early or pre-attentive vision so that there has not been a serious attempt to define and/or come up with a general representational scheme for textures. The few that have been proposed (e.g. [12, 13]) have not proven adequate.

There is another reason for being interested in the description of texture. The process of description illustrates how one can build a representation and description scheme for images/pictures of arbitrary complexity. A picture is supposed to be worth a thousand words. However, most people would be hard pressed to come up with the thousand words that describe a picture! This is a serious problem if the world moves towards pictorial information through graphics and visualization. What good is it to represent information pictorially if we are not comfortable with analysing and understanding the resulting images?

We focus on a single visual cue to illustrate this problem. In the case of texture, we find that our ability to explicitly analyze and represent texture is very limited. Given the increasing importance of image representation and analysis in fields ranging from art to visualization, we feel that this is a serious bottleneck in understanding the content of images. We take an interdisciplinary approach toward the solution of various problems raised by a consideration of the property of texture. In order to overcome the deficiency in representing textures, we present a description scheme for textures that builds upon the work of Rao [12]. Briefly, we categorize textures along the qualitative dimensions of repetitive vs. random, oriented vs. non-oriented, and simple vs. complex. These qualitative dimensions have been shown to be used by human subjects in texture discrimination tasks by employing techniques from psychology [14, 15]. Thus, these categories appear to be "natural" for human subjects in texture perception and description.

Our approach, which combines philosophical insights into object representation, psychological studies on subjects, and quantitative techniques from computer texture analysis, paves the way to a standardized representation for texture. We hope that this will be a foundation for a Texture Naming System, much like the Color Naming System in computer graphics [16].

In this paper our plan is to:

1. Locate texture in the context of the philosophical distinction between primary and secondary properties.
2. Present a proposal for a qualitative representational scheme for visual textures.
3. Discuss the relevance of results of two categorization tasks performed by human subjects (a) How subjects categorize textured *images* (categorization in the visual domain); (b) how subjects categorize texture *words* (categorization in the verbal domain).
4. Extract conclusions and suggest some directions for future work.

11.2 Benefits of This Research

One of the outcomes of the analysis presented in this paper will be to provide a basis for understanding the symbolic or verbal description of texture. This problem of signal-to-symbol transformation [17, 18] is one of the central issues in computer vision. The problem is usually tackled by first computing features of the image and then classifying the image (or portions thereof) into a set of pre-defined classes.

For many visual cues, it is possible to define these features precisely, and we also know the dimensionality of these features. For instance, color is characterized in terms of brightness, hue, and saturation of the RGB components [19]. Due to this three dimensional nature of color, traditional color specification systems use a triple of numbers to describe color [20, 21].

Unfortunately, such a parallel does not exist for texture since we do not have a firm handle on the dimensions of texture. In an attempt to understand the semantic significance of texture, Rao [12] devised a taxonomy for texture, based on an analysis of a variety of textures. However, the proposed taxonomy was derived from purely computational considerations, and was not backed by any psychophysical evidence. The analysis presented in this paper will shed new light on the categorization of texture.

The benefits of conducting this research are:

1. It enables a better characterization of texture features, and suggests the salient features used for texture perception.

2. It suggests ways of classifying textures, and creating a taxonomy for texture [12]. Such a taxonomy could be used by inspection systems for creating a defect classification scheme [22, 23].

3. It suggests the features that need to be extracted for computational texture analysis and classification.

4. Through an understanding of the taxonomic relationships among a broad range of textures, we hope to help users in graphics and visualization as well as computer vision construct effective algorithms for texture rendering and recognition.

11.3 Primary versus Secondary Properties of Objects

We will take the Lockean distinction to be characterized by the following differences, which are interrelated, between primary and secondary properties:

1. Primary properties are those that one can attribute to an object at *any* structural level, whether microstructural or macrostructural. They are therefore 'universal' properties. These are the properties of objects with which physics concerns itself.

In contrast, secondary properties 'emerge', as it were, only at higher or macro levels, i.e., the level at which the sensory mechanism of the particular type of viewing organism is able to react to light and other waves as they reflect or refract from the object. Of course, the kind of sensory mechanism invoked is biochemical when it comes to taste or smell. The molecular properties of an object of a certain shape, size and polarity triggers the appropriate taste sensor (sweet, sour) which in turn gives rise to tastes. We are concentrating on vision in this paper and so we in turn concentrate on the physical properties of objects as they are affected by light and other waves. These are properties that may be the concern of 'higher level' sciences such as biophysics, neurophysiology, biology, or psychology.

2. Primary properties are those that an object could have *independently* of the particular visual or sensory mechanism of the viewing organism. Secondary properties are *dependent* properties, i.e., they depend, for their existence, upon the interaction between the object (in virtue of its primary properties) and the sensory mechanism of the subject.

3. Primary properties are *intrinsic* properties of an object. They inhere in the object autonomously, i.e., independent of and uncaused by the properties of any other object. Secondary properties are *extrinsic*; they do not inhere in the object at all but are caused in part by the configurations of the inherent primary properties of the object in question and in part by the configuration of the viewing mehanism.

For Locke the three differences are not mutually exclusive but are interrelated. On this view, primary properties attach to an object at any level of structural resolution (or complexity). This suggests that they are fundamental to objects. They are independent properties: independent, on the one hand, of the properties of the viewing mechanism; on the other, of properties of other (non-viewing) objects. The former kind of independence focuses on the independence of primary properties from influence of properties of *a specific kind* of other object: sentient physical objects. The latter expresses a more radical kind of independence, where primary properties of an object are regarded as independent of the properties of all and any other physical objects. This is captured by the feature of inhering in, being intrinsic to the object-in-itself.[5]

What is the upshot of granting this sort of distinction between types of properties? For one, it forces a commitment to a specific conception of an object. As we follow Locke, primary properties attach to a physical object at all levels. Could an object fail to have a primary property and still *be* an object? To answer this question, let us go to Locke. Consider, for instance, what Locke has in mind when he calls primary properties 'constant'. These are not properties that are lost as we move to finer-grained structural levels or which emerge at higher or coarser-grained structural levels. Rather, they are found "in every particle of Matter" (Book II, Chapter viii, section 9, pp. 134-135) [2]. This is not simply based on what our Senses tell us regarding the nature of these properties, revealed in the constancy of the ideas we have of these properties as we view the object. That is, it is not evidence we gather based on actual observation. Rather, it is based on what the Mind tells us: for this is what "..the Mind finds inseparable from every particle of Matter.." (p.135). This suggests that Locke is not making an empirical point here, but rather a conceptual point about the way to think of the connection between an object and its primary properties.[6]

Primary properties of an object, then, are properties that an object has essentially, in virtue of which it may be called an *empirical or physical* object. What this means is that it makes no sense to conceive of an *object* that at some level of resolution might lack one or more of the primary properties; i.e., that may lack figure or number or solidity.[7] So if it turned out to lack one of these properties at a certain level, we could on Lockean grounds deny to it the status of being an

---

[5]Locke's own view is that primary properties inhere in a 'substance', which itself has no properties of its own, but is simply and fundamentally 'a something-I-know-not-what' and therefore distinct from the empirical object that is the focus of study by physicists. Opinion is divided on how to understand Locke's doctrine of substance. Some interpret substance as an essence that is in principle unknowable; others that it requires scientific inquiries that are in fact beyond our capacities and therefore that substance is in fact though not in principle unknowable. See, for instance, Gibson [24]. Bennett [3] shares Gibson's view on this.

[6]Margaret Wilson notes the different interpretations of Locke on this issue. Thus she reads Bennett as imputing the empirical point to Locke, with which she disagrees, while others have taken Locke to be making a theoretical point here [25].

[7]In contemporary times we would replace solidity with the property of mass.

empirical object. The concept of all and only color (like phenomenal redness, without boundaries or size or texture or..) would not be the concept of an empirically robust object. Also, this is a conceptual point, not an epistemological point that has to do with the character of our recognition of these properties. This is therefore consistent with the fact that our *perception* of primary properties like shape and size is dependent upon extrinsic factors in that it depends at the macrostructural level upon how organisms view the object. Thus shape and size are not *inherently* dependent upon extrinsic factors in the sense that objects are thought to be shaped and sized quite independent of choice of instrument or an organism's capacities for discovery. That an object has shape or size is an intrinsic fact about our concept of an object. Another way of putting the point is to say that our *concept* of an object's shape or size is delineated in a way that excludes the relationship to a viewing organism of any kind. The latter relation is an extrinsic one.

In contrast, properties like color, smell, taste, sound, and feel are intrinsically rather than extrinsically related to the capacities of the viewing organism and the presence of light and other physical mediums. The redness of the apple is a property that is *constituted* by the particular wave length to which my eyes are sensitive. The apple would not have the property of being red in itself. Thus these properties of objects are constituted by properties of entities outside the parameters of the object. This is also part of our understanding of the *concept* of color.[8]

For this reason, these properties are viewed by Locke as *relational*, while properties like shape and size are considered *non-relational*. Of course, primary properties are 'dependent' properties in the minimal sense that they do depend upon not only choice of measuring instruments but upon frame of reference. Still, there is a difference. They are related to other entities in the physical world, not to this par-

[8]One may distinguish between concept and property. In a quite different context, Jaegwon Kim proposes a distinction of this kind [26]. He suggests that the concept of belief may be distinct from the property of belief and so for all mental concepts. Thus a single concept may map onto quite distinct physical properties. We may be read as making an analogous point; that at the very least there is a distinction to be made between the concepts of color and shape, which are quite distinct, and the actual properties of color and shape, which may be empirically (or physically) resolved in quite similar ways. Thus distinct types of concepts may map onto quite similar types of physical properties.

ticular physical object, namely, the sentient viewing organism. It is this distinction that we intend to capture by our characterization: consitutively relative-to-sentient-viewing-organism (secondary); not (primary).

11.3.1 Problems and Clarifications

Locke and his contemporaries did not have to deal with the different kinds of 'objects' that have become commonplace with the advent of technology. Some of these do not have many of the primary properties thought to be essential to an object. Computer images are an example. Here clearly it is not the case that a computer image has to have all the primary properties like mass (controversial even as we get to the quantum level) or weight or density (except in terms of its visual surface properties, eg. looks heavy, looks dense) in order to be a perfectly respectable object.[9] However, one can make sense of Locke's view, i.e., of maintaining a strong connection between an object and its primary properties, by distinguishing between empirically robust objects and thinly empirical objects such as computer images and, perhaps, holograms. Certainly objects such as these, and, more recently, objects in virtual or cyber space put pressure on our thinking of primary properties as necessary to objects.[10]

In addition to the distinction between primary and secondary properties, we suggest a second distinction between the objective and the non-objective. This we will take to be a distinction between what 'can be quantified' versus 'cannot be quantified', or, simply: quantitative versus qualitative. Primary properties like weight and size are clearly quantifiable properties of an object. Once you've picked out your object, you can compute its weight, size, density etc. However, color has spectral properties. This means that although we must identify color phenomenally as 'redness' (how it appears to us), it can also be measured physically in terms of wave lengths. So color is quantifiable in this sense, as are size and weight. The question of

[9] See Smith [4] for a development of the suggestion that the only true contenders for being primary are the properties, respectively, of being extended in space and in time. This would allow one to leave out the problematic properties of mass and weight, for example, as necessary to the concept of an object.

[10] Armstrong [27] questions whether the list of primary qualities *suffice* to give us a physical object; curiously enough, he does not raise the issue of whether or not they are *necessary* to objects. He simply assumes that they are.

quantifiability, then, cuts across the primary/secondary property distinction. While the latter distinction is conceptual and in this sense non-contingent, we take the distinction between what can and what cannot be quantified to be a contingent distinction, dependent upon the technological discoveries of the time. Thus it is conceivable, for instance, that taste, although secondary and currently non-objective in our sense (we cannot at present quantify over tastes), could become objective as research and technology into the physiology of taste develop and where the verbal descriptions have been systematized sufficient to allow for the construction of algorithms that would map those qualititative descriptions in a quantitative way.

In conclusion, we retain the distinction between primary and secondary properties, but introduce an additional distinction between the qualitative and the quantitative which is of particular significance for researchers in the sciences in the categorization and analysis of objects and their properties.[11]

11.4 How to Think About Texture

Into which categories does texture fall? The debate about primary and secondary properties raises an issue about a distinction between deep-structure (think 'objective') properties and surface (think 'accessible-to-viewer') properties that bears more directly on our discussion of texture. Smith [4] argues for precisely this way of understanding the 17th century distinction between primary and secondary properties. Thus, according to him, the primary properties were those that belonged to the physical fabric of the world while the secondary properties were those that were 'peculiarly sensory', i.e., those that our senses made us believe were in the object. In this paper we are interested in texture understood as a property of surfaces rather than as one of deep structure, although, as we shall see, there is a way of understanding texture that puts it into the category of deep-structure.

For Locke, texture is a property that combines shape, size, and density of the particles (elements) that make up the object config-

[11]Smith [4] correctly criticizes Australian philosophers Armstrong [6] and Campbell [9] for conflating the distinction between primary and secondary properties and between properties that can and cannot be scientifically measured. The issue of quantifiability is not what makes a property primary.

uration. On this understanding of texture, it is distinct from the standard primary properties in that it is complex. That is, a complex of certain primary properties yields an object's texture. Thus, as Smith [4] points out, it is unique in the Lockean scheme of properties in that it is the only complex property that "is explicitly defined in terms of the primaries". All other complex properties such as being malleable or magnetic are simply "determinate configurations that matter, as defined by the primaries, can take up". In other words, these so-called "properties" are simply alternate ways that the primary properties such as shape, weight, density and so on come together in a particular object. Texture, in contrast, "is that *emergent* spatial configuration that results when two or more atoms are conjoined to form a body" (p. 234, emphasis added).

An area of research which might serve to provide us with a good example of the 'emergent' complex property of texture, understood in Locke's sense, is crystallography. Researchers from areas as diverse as mathematics, chemistry and art are interested in the inner structure of the crystal mostly because they are interested in symmetry, and crystals are known for their symmetrical structure [29]. The goal is to find out the internal structure of a crystal or the orderly arrangement of molecules within a crystal by shining an X-ray through it; the symmetry of the diffraction pattern reflects the symmetry of the internal atomic ordered structure of the crystal. There are a limited number of diffraction patterns for a crystal. If a compound has a disordered structure the diffraction pattern of the X-ray exhibits disorder and is unable to reflect accurately the structure of that compound. It appears then that the underlying 'texture' of a crystal can be described in quantitative terms as instantiated by our ability to compute the X-ray's diffraction pattern. We can view this as a good example of how we may quantitatively represent texture at least with respect to compounds that exhibit order. In addition, it appears at the level of deep structure and is independent of the viewer. It therefore satisfies two of the three characteristics for primariness. Finally, it is not universal, for a single atom in the crystal has (no structure, and hence) no texture. This would seem to put texture into the category of the 'complex primary', just as Locke had suggested.

In this example, however, the 'texture' identified here is, in effect, simply a label for a property that may be described in terms of more fundamental properties (i.e., shape, size and the density of atoms). That is, the property we identify as texture is identical, both conceptually and materially, with the property of orderly arrangement of atoms at this level, and therefore is not an 'emergent' property in any real sense, as Locke seems to have thought.

We suggest, alternatively, that it is at the level of human perception rather than at the level of deep structure that texture 'emerges' as an interesting and useful property in its own right. For instance, when one describes the pattern of a carpet, or the orange-peel appearance of a silicon wafer chip, or the swirls in a knot of wood, one is focusing upon the object's visual *texture*. Thus we propose an alternative definition of texture as the surface markings on an object or the 2-D appearance of a surface (e.g. reptile skin, wood, carpet, etc).

11.4.1 Texture as a Property of Visual Surfaces: Secondary and Non-objective (Qualitative)

We may informally think of texture as the 'pattern' of an object. The texture we see at the macroscopic level may or may not correspond to the texture of the microstructure of the object. A wood table may have a knotty, swirly texture at the visual level but a linear, evenly granular texture at the non-visual level. At this level of analysis the 'texture' of an object is difficult to quantify. One reason is that at this level we have the kind of instrument that would permit this as we also have at several sub-visual levels where molecular structure may be measured by X-rays and electron tunneling microscopes. Neither of these instruments tell us anything about the texture of the *visual* surface. Our unaided eyes are of course the obvious 'instruments' at the visual level but there is no clear way of quantifying what texture we see *directly*, by simply recording what we see. Simply reading off information contained on the retina, for instance, is of no help to us, because at the visual level 'what we see' is inextricably dependent upon our purposes as we delineate the objects for ourselves. And this fact is even more salient when it comes to picking out the patterns or textures in the objects that we view, quite apart from the more simple properties like only shapes or only colors[12] So, not

[12]We should hasten to add that we do not mean to exclude color and shape from a possible characterization of an object's texture. Indeed, sometimes the

surprisingly, the discussions of visual texture have for the most part proceeded at the *qualitative or descriptive* level.

Indeed, in the area of visual textures the more fruitful approach is to proceed at the *qualitative* level, to come up first with a qualitative description scheme for textures. Here is where our work can provide some suggestions. Texture as a *qualitative* visual property is critical for performing many tasks. We need first to fix on the visual pattern initially qualitatively or descriptively. In other words, there is no objective or non-descriptive starting point to gain access to the kind of texture described by the descriptions. So we have to begin with the descriptions. From this the task is whether the qualitative categorization may be quantified so that a machine could use the human qualitative descriptive categories for texture to distinguish mechanically one kind of texture from another the way a human would.

As a concrete example, consider the area of semiconductor wafer inspection, where it is common to find qualitative descriptors such as hillocks, grass, worm-hole, starburst, orange peel, and speedboat. This jargon is popularly used to describe different kinds of anomalies and defects arising during wafer processing. There are few quantitative measures for these kinds of properties. For instance, the Hewlett Packard Photolithography Advisor [30] defines an *orange peel defect* to occur when 'the photoresist has a wrinkling effect similar to the skin of an orange.' The system also asks the user to make a distinction between light and heavy orange peel – which is very subjective.

Such a scheme is ad-hoc, and a more scientific scheme is desirable. Such jargon becomes necessary due to the lack of a standardized vocabulary for texture description. We will address this standardization issue later in this paper.

At the visual level, texture continues to be understood and viewed as a global property of the object where the size and shape of 'visually-atomic' particles along with their 'density' or degree of 'packedness' yields the object's texture. At this level, however, in contrast to Locke's thoughts about texture, the choices of the viewer come into play at all times, and in this sense it is always viewer-centered.

---

texture of an object may be captured solely by discontinuities in its colors. Our point is rather that shape of an object or color of an object is a simple property that may be more readily quantifiable than the complex property of texture.

11.5 Understanding Features Humans Use in Texture Perception

Texture is a property that can be perceived either visually or through touch. We need to make a distinction between the visual and tactile aspects of texture. Though this paper will deal primarily with the visual aspect of texture, we briefly mention some interesting studies on tactile texture. Yoshida [31] used a multidimensional scaling approach to identify the principal dimensions of touch. This study found the main dimension to be metallicness *vs.* fibrelikeness. The physical dimensions that differentiate these two extremes are specific gravity, thermal conductivity, plasticity, and hardness. Ohno [32] investigated the visual perception of surface roughness of building materials. Subjects used six bipolar scales (rich *vs.* poor; rough *vs.* smooth; soft *vs.* hard; warm *vs.* cold; light *vs.* heavy; bright *vs.* dark) to judge building materials. A significantly high correlation was noticed between hard-soft, heavy-light, and warm-cold.

Psychophysicists have studied texture because it provides them with an insight into early human visual information processing. Most psychophysical studies on texture perception have focussed on the *preattentive* aspect of perception, which refers to the fact that the cognitive process operates over a very short duration, and does not involve concentrated attention on the stimulus. The subject is typically required to perform a discrimination task after being presented with a pair of images for a duration of 100 milliseconds [11]. The human visual system is capable of discriminating between certain kinds of textures preattentively. Psychologists have considered this ability to be based on the preattentive detection and processing of certain primitive features. Some of the features that have been identified are elongated blobs (rectangles, ellipses, or line segments) with specific properties such as color, orientation, width and length; the ends of lines (terminators) and crossings of line segments.

A noteworthy aspect of texture perception is that the *discrimination* between textures is a much simpler task than the *identification* of textures. Thus it is far easier for a subject to decide if two textures are similar or dissimilar rather than to describe those textures. Nevertheless, there are many tasks that require the *attentive* analysis of textures. For instance, an artist examining the texture of Van Gogh's brush strokes, or an operator inspecting the surface texture of a finished product engage in a deliberate analysis of texture. Hence it is of interest to determine how humans classify textures attentively, what features of textures are used for discrimination, and what the organization of the texture feature space is. In order to motivate this approach to the problem, it is illustrative to consider the research done in investigating other cues such as color.

11.5.1 A Comparison With Color

It is interesting to compare texture and color from a human perception as well as a computational viewpoint. We believe that a lot of interesting research has been conducted in the domain of color, and studies in texture can benefit from the approaches and ideas used in understanding color.

For instance, the Dictionary of Color [33] was assembled to serve as a common ground for the proper appreciation of all existing terms and the specific colors they represent. A significant amount of collation had to be done, drawing on color words from diverse sources such as the arts, sciences, and industries. The resulting dictionary has over 2000 words on color! Interestingly, the authors of this dictionary, Maerz and Paul, point out that "since the use of color in its brightest aspects has grown common in everyday life, in dress, printing, interior decoration, and the growing tendency to use color for exterior architectural embellishment, the need has become urgent for standardization and for a complete reference source of color names." It would be interesting to carry out such task in the case of texture.

Color can be characterized by a variety of three-dimensional representations (HLS/RGB, etc). However, no comparable scheme for texture exists. We believe that part of this is due to the fact that a standardized taxonomy for texture does not exist, and the dimensions of texture have not been rigorously identified.

The easy specification of color has been made possible for users of computer systems through the Color Naming System (CNS) [16]. A parallel need is felt for a Texture Naming System, which would standardize the description and representation of texture. The standardization of vocabularies for features such as color, shape, and texture would be useful for various kinds of applications, ranging from graphics to automatic defect classification [34]. In the case of color, the Color Naming System (CNS) [16] was set up to fulfill the need for standardizing linguistic descriptions of color. The advantage of the CNS is that it achieves the goal of standardization by using simple, easily understood primitives from the English language. The CNS quantizes the three dimensions of color, viz. hue, lightness, and saturation, into three discrete sets of symbols. Thus, lightness has the five discrete values of very dark, dark, medium, light, and very light; saturation has four values: grayish, moderate, strong, and vivid; hue can take the values black, very dark gray, dark gray, gray, light gray, very light gray, white, blue, purple, red, orange, brown, yellow, and green. Further modifications of hue names are possible through the use of the suffix "ish", e.g. greenish blue. A grammar specifies how these words may be combined. Examples include colors such as light greenish blue, medium strong red, etc.

A noteworthy feature of the CNS is that it simplifies and systematizes the Inter-Society Color Council-National Bureau of Standards Lexicon of Color, containing names for 267 regions of color space. This lexicon is based on earlier dictionaries of color and on a detailed understanding of human color perception.

We believe that similar efforts to standardize descriptions of textures, grounded in human perception, should be undertaken. This will result in fairly generic features which can be expected to work in a wide set of domains as well as a lexicon for naming textures.

11.5.2 Computer Perception of Texture

It is instructive to first consider the vast body of literature on computer texture perception, partially reviewed in [12]. Most of the computer texture analysis techniques are based purely on computational considerations, and are not modeled after the human visual system. There are three broad classes of texture for which computational techniques exist: structural, oriented, and statistical. Structural textures (e.g. brick wall) can be described in terms of a primitive element that has been repeated in a plane according to certain placement rules [35, 36]. Oriented textures (e.g. wood grains) are characterized by an orientation field, which assigns a local direction and strength of directionality to each point in the texture [37]. Statistical textures (e.g sand) are those that show neither repetitiveness nor orientation. They are usually characterized in terms of measures such as fractal dimension [38] or the co-occurrence matrix [13]. Many of these computational measures for textural features are used for scene segmentation [39] or for classification [40].

Interestingly, these three categories emerged in a *de facto* manner, and there was no theoretical analysis to show that these categories were sufficient to represent all visible textures. Fortuitously, these three categories also emerged as being significant from a study of human texture perception, described in the next section. In other words, the categories of algorithms chosen by computer vision researchers have a definite correspondence with the categories of texture perceived by humans.

11.5.3 Identifiying the Dimensions of Visual Texture

Earlier work on texture classification schemes, e.g. Rao [12] was based on intuition and computational considerations, and has not been adequately verified by psychophysical studies. While these intuitions have yielded some insights, empirical work is required to discover and elaborate the basis on which people perceive texture. We need to understand how people classify texture into meaningful, hierarchically structured categories.

In this section we review two experiments aimed at understanding the dimensions of visual texture. In the first experiment, Rao and Lohse [14] used a multidimensional scaling approach to unravel the structure of the texture feature space. Multidimensional scaling has been used successfully in other disciplines for similar purposes, and was well suited for this task. Subjects were asked to group a set of 30 visual textures from Brodatz's album [41] into different categories. Subjects were free to choose their own categories. Data from such an unsupervised classification was used, and subject to multidimensional scaling and hierarchical cluster analysis. Three important dimensions for texture perception were tentatively identified viz. repetitiveness vs. irregularity, directional vs. non-directional and structurally complex vs. simple.

This study was lacking in that it did not use any quantitative data from the subjects: only non-metric grouping data was used. As a consequence, it is difficult to answer questions about which dimension is more important, or what is the relative importance of these dimensions. In order to properly understand these metric issues, relevant dimensions hypothesized by the first experiment were used to construct metric scales for the second experiment [42]. These metric scales were then evaluated by 20 subjects on 56 textures from

Brodatz's album [41].

This data gathered in the second experiment [42] confirmed the basic categories from the first investigation and enabled a classification of visual texture. The results identify the attributes that people may use to judge similarity among visual textures. The orthogonal dimensions identified were repetitive *vs.* non-repetitive; high-contrast and non-directional *vs.* low-contrast and directional; granular, coarse and low-complexity *vs.* non-granular, fine and high-complexity. This interpretation fits the data well, and refines the results we presented earlier [14]. Figure 11.2 summarizes the orthogonal dimensions of texture that were uncovered in this study [42].

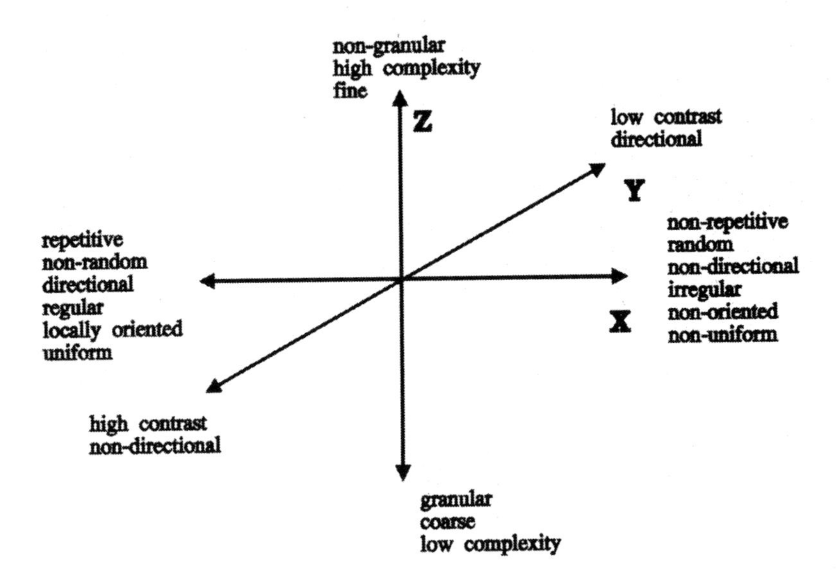

FIGURE 11.2. The three dimensions of texture.

11.6 Experiments on the Categorization of Texture Words

The results from the above experiments have led us to the third experiment, which investigates the verbal categorization of texture: given words in the English language which deal with texture, what categories do people form with these words.

The results from our earlier experiment on categorization of visual texture showed that there is close similarity in the way people group pictures of texture. This led us to wonder whether people would group just the words related to texture in a similar fashion. If there is indeed some similarity, then a suitable symbolic representation scheme for texture can be designed, which has human texture perception and description as its basis.

We performed a search of words in the English language that dealt with surface texture properties, using Webster's on-line dictionary. Our search resulted in around 300 words. We pared this original list down to 98 by deleting the less frequently used words. Based on a pilot study conducted at Smith College, we realized that subjects had a hard time categorizing more than 100 words, so we decided to keep the categorizing task amenable.

Subjects were given the 98 words on cards and asked to categorize them by first visualizing the look of a surface that they think is captured by the word – eg. visualize the look of a surface that is blotchy or grilled or peppered – and to group together the words that seem to them to be similar in terms of the look. We are in the process of analyzing data gathered from 40 subjects. This is an ongoing study, and the results will be reported shortly [15].

We present some preliminary results gathered from hierarchical cluster analysis. We found the most common groupings used by the subjects to be

1. lined, grooved, ridged, furrowed, pleated, corrugated, ribbed, striated, stratified, banded, zigzag.
2. matted, fibrous, knitted, woven, meshed, netlike, cross-hatched, chequered, grid, honeycombed, waffled.
3. flowing, whirly, swirly, winding, corkscrew, spiraled, coiled, twisted.
4. faceted, crystalline, lattice, regular, repetitive, periodic, rhythmic, harmonious, well-ordered, cyclical, simple, uniform, fine, smooth.
5. complex, messy, random, disordered, jumbled, scrambled, discontinuous, indefinite, asymmetrical, non-uniform, regular.
6. spattered, sprinkled, freckled, speckled, flecked, spotted, polka-dotted, dotted.

7. gauzy, cobweb, webbed, interleaved, entwined, intertwined, braided, frilly, lacelike.

8. mottled, blemished, blotchy, smeared, smudged, stained.

9. wizened, crows-feet, rumpled, wrinkled, crinkled, cracked, fractured.

10. marbled, veined, scaly.

11. bubbly, bumpy, studded, porous, potholed, pitted, holey, perforated, gouged.

Further analysis that will be performed includes multidimensional scaling and principal components analysis.

We propose to combine the results of these experiments to create a Texture Naming System, which specifies the dimensions of texture along which variations can be described.

11.7 Conclusion

Locke recognized quite rightly that texture was a unique property of objects. However, he misidentified its nature, opting to classify it as a complex property more akin to being primary than secondary. We argue instead that texture is an emergent complex property of surfaces, and secondary. We provide the beginnings of an analysis of visual texture that builds on earlier work of researchers in computer vision and distills the results of two psychological experiments – on textured images and on texture words. Preliminary results suggest that three dimensions of texture are fundamental: repetition, orientation, and complexity.

We propose to combine the results of these experiments to create a Texture Naming System, which will form a standardized natural language interface for the description of visual texture. From a broader viewpoint, our approach helps in understanding how to build a visual language, as we address the fundamental question of what the elements of images are, and how they are composed.

The interdisciplinary approach to this problem is crucial to the success of our work: we have combined philosophical insights, psychological studies, and ideas from computer graphics and vision.

11.8 References

[1] Boyle, R. *Works*, vol. iii.

[2] Locke, J. (1975). *An Essay Concerning Human Understanding.* Oxford University Press.

[3] Bennett, J. (1971). *Locke, Berkeley, Hume: Central Themes.* Oxford: Clarendon Press.

[4] Smith, A. (1990). Of primary and secondary qualities. *Philosophical Review,* XCIX (2) , 221–254.

[5] Hardin, C. (1988). *Color for Philosophers: Unweaving the Rainbow.* Indianapolis: Hackett.

[6] Armstrong, D. (1981). *The Nature of Mind and Other Essays.* Ithaca: Cornell University Press

[7] Boghossian, P., and Velleman, D. (1989). Color as a secondary quality. *Mind,* 98, 81–103.

[8] Boghossian, P., and Velleman, D. (1991). Physicalist theories of color. *Philosophical Review,* 100, 67–106.

[9] Tolliver, J. (1992). Papers on color. In *Philosophical Studies.* Dordrecht: Kluwer Academic Publishers.

[10] Dennett, D. (1991). *Consciousness Explained.* Boston: Little, Brown and Company.

[11] Julesz, B. (1981). Textons, the elements of texture perception and their interactions. *Nature,* 290, (March), 1619–1645.

[12] Rao, A. R. (1990). *A taxonomy for texture description and identification.* New York: Springer-Verlag.

[13] Haralick, R. M. (1979) Statistical and structural approaches to texture. *Proceedings of the IEEE,* 67 (5), (May), 786–804.

[14] Rao, A., and Lohse, G. (1993). Identifying high level features of texture perception. *CVGIP: Graphical Models and Image Processing,* 55 (3), (May), 218–233.

[15] Rao, A., Bhushan, N., and Lohse, G. Experiments in texture naming: categorization of texture words in the English language. In press.

[16] Berk, T., Brownston, L., and Kaufman, A. (1982). A new color naming system for graphics languages. *IEEE Computer Graphics and Applications,*2 (3), (May), 37–44.

[17] Hanson, A., and Riseman, E., Eds. (1979). *Computer Vision Systems.* Academic Press.

[18] Rao, A. R., and Jain, R. (1988). Knowledge representation and control in computer vision systems. *IEEE Expert,* Spring, 64–79.

[19] Wyszecki, G., and Stiles, W. S. (1967). *Color Science.* New York: Wiley.

[20] Hearn, D., and Baker, M. P. (1986). *Computer Graphics.* Prentice Hall.

[21] Farhoosh, H., and Schrack, G. (1986). CNS-HLS mapping using fuzzy sets. *IEEE CGA,* 6 (6), 28–35.

[22] Godinez, P. A. (1987) Inspection of surface flaws and textures. *Sensors,* (June), 27–32.

[23] Rao, A. R., and Jain, R. (1990). A classification scheme for visual defects arising in semiconductor wafer inspection. *Journal of Crystal Growth,* 103, 398–406.

[24] Gibson, J. (1960). *Locke's Theory of Knowledge and its Historical Relations.* Cambridge.

[25] Wilson, M. (1992). History of philosophy in philosophy today; and the case of the sensible qualities. *The Philosophical Review,* 101.

[26] Kim, J. (1992). Multiple realization and metaphysics of reduction. *Philosophy and Phenomenological Research,* LII, 1 (March).

[27] Armstrong, D. (1961). *Perception and the Physical World.* London: Routledge and Kegan Paul.

[28] Campbell, K. (1972). Primary and secondary qualities. *Canadian Journal of Philosophy* (72).

[29] Senechal, M., and Fleck, G., Eds. (1977). *Patterns of Symmetry.* Amherst: University of Massachusetts Press.

[30] Fong, W., Cline, T., Walker, M., and Rosenberg, S. (1985). A photolithography advisor: an expert system. In *Proc. SEMICON/EAST 85*, pp. 1–8.

[31] Yoshida, M. (1968). Dimensions of tactual inpressions. *Japanese Psychological Research,* 10 (3), 123–137.

[32] Ohno, R. (1980) Visual perception of texture: development of a scale of the perceived surface roughness of building materials. *Environmental Design Research Association*, 11, 193–200.

[33] Maerz, A., and Paul, M. (1950). *A Dictionary of Color.* New York: McGraw Hill.

[34] Chou, P., Rao, A.R., Sturzenbecker, M., and Brecher, V.H. (1993). Automatic classification of defects for wafer inspection. In *SPIE Conference on Machine Vision Applications in Industrial Inspection,*1907, (February).

[35] Hamey, L. G. C. (1988) *Computer Perception of Repetitive Textures.* PhD thesis, Computer Science Department, Carnegie Mellon University, Pittsburgh.

[36] Tomita, F., Shirai, Y., and Tsuji, S. (1982) Description of textures by a structural analysis. *IEEE Trans. Pattern Analysis and Machine Intelligence,*4 (2), 183–191.

[37] Rao, A. R., and Schunck, B. G. (1991). Computing oriented texture fields. *CVGIP: Graphical Models and Image Processing,* 53 (2), (March), 157–185.

[38] Yokoya, N., Yamamoto, K., and Funakubo, N. (1989). Fractal-based analysis and interpolation of 3d natural surface shapes and their application to terrain modeling. *CVGIP,* 43 (3), (June), 284–302.

[39] Laws, K. I. (1980). *Textured Image Segmentation.* PhD thesis, Dept. of Electrical Engineering, Univ. Southern California.

[40] Gotlieb, C., and Kreyszig, H. (1990). Texture descriptors based on cooccurrence matrices. *CVGIP*, 51 (1), 70–86.

[41] Brodatz, P. (1966). *Textures: A Photographic Album for Artists and Designers.* New York: Dover Publications.

[42] Rao, A., and Lohse, G. (1993). Towards a texture naming system: Identifying relevant dimensions of texture. *IEEE Conf. on Visualization,* (October). To be published.

12

Variability and Universality in Human Image Processing

Presenting A Model Of Human Perception, Symbolic Representation, Meaning Derivation, Communication And Transformation

Beverly J. Jones[1]

12.1 Introduction

At the first IEEE Conference on Visualization in 1990 Dr. James M Coggins stated, "Closing the loop between image understanding and image generation is a natural and necessary step in creating interactive visualization environments...."([1], p. 408). He provided a list of eighteen concepts that students in any image field should understand. These are heavily oriented toward a machine oriented technical perspective although he does include four that are human oriented, i.e., models of human spatial vision, models of human color vision, visual illusions, and stereopsis. While all of the items on his list are important, his technologically oriented perspective may be augmented through emphasis on human image understanding and generation regarding the machine as an extension of the human. From this perspective most important technical problems are actually human conceptual problems. Exploring underlying conceptual frameworks may assist in a reassessment of assumptions.

This paper presents a model of factors influencing perception, visual information processing, cognition, and culture. This model draws upon research in the arts, cognitive sciences, neurosciences, psychology, anthropology, sociology, education, and computer science. This model stresses the existence of both variability and universality in biological, psychological, and socio-historical contexts. It

[1]Department of Art Education, School of Architecture and Allied Arts, University of Oregon, Eugene, OR 97403-1206

has evolved over the last twenty years from the study of human visual perception, imagery, understanding, as well as creation, presentation and evaluation of symbolic and material images and representations. Within the limited scope of this paper primary points related to the model and imagery, particularly visual imagery are discussed. Finally, implications are derived from these points for future study.

12.2 Background

A model of human understanding of images is necessarily multi-faceted, complex and draws upon multiple disciplines. Such a model is necessarily shaped by the culture in which it is created. An examination of the limits of prior models, especially those of models built upon nineteenth century modern thought, may prove useful to computer scientists interested in construction of graphic interfaces, computer graphic models of nature, and of symbolic knowledge and similar topics. This paper draws mainly on the interrelationships of two areas, computer graphics and artificial intelligence models combining these with conceptual concerns expressed in other disciplines regarding human imagery.

In a series of recent papers [2, 3, 4, 5], the author has advanced the following perspective: Conventions reflecting larger models of humanly constructed cultural and historical reality are embedded in imagery, including electronic imagery created for aesthetic/artistic or scientific/technical purposes. Careful analyses of the form, content, and practices surrounding this imagery reveals these views of reality and those embedded in models generated using this technology. The interrelationships of technological developments, cognitive science, understanding imagery, learning, and art education have been stressed in another paper [6].

These articles and the model presented in this paper assume that the form and content of information and the acquisition, maintenance and utilization of information are shaped by complex interacting variables including factors related to individual human participant (intra and inter individual factors), the immediate context (pyscho-cultural, visual-physical, etc.), and the larger context (historical, cultural, physical, etc.)

Dimensions stressed in the three part model of human image processing presented in this paper include:

- Factors in the shaping of perception, imagination and understanding of imagery. These include interior factors, sensory interface factors, and immediate and larger environmental factors.
- Factors in the shaping of communication, expression, delineation, presentation of imagery, and its evaluation and potential for transfer.
- Factors involved in transformative nature of imagery beyond communication. These factors are related to the way imagery affects cognition, cognition affects imagery production, and other sometimes unexpected culture change.

These three parts are viewed as dynamic and interactive. They are considered for universality and variability as seen through the four filters in Figure 12.1. These filters result from structures of thought existing simultaneously in the contemporary world. The interactive, dynamic, and self constructive nature of human perception; experience of imagery; and the contribution of imagery to the shaping of larger reality constructs is stressed in the model I am presenting. From this perspective a broad view of imagery, cognition, and culture change is presented.

Represented in Figure 12.1 are four simultaneously existing contemporary views. These may be related to imagery, representations, and underlying imagistic constructs that shape knowledge structures in our culture. This paper is necessarily self-reflexive in that it takes one of the views (View Four) as a conceptual model for the paper while using View Two as a presentation model. The conceptual model is heavily influenced by computer graphic imagery of natural phenomena. An earlier paper [5] proposes that this imagistic construct is contributing to the breakdown of the dominant thought paradigm of our culture that emphasizes universality (View Two). It also negates the completely relativist tendencies of View Three that illustrates a reaction to the dominant paradigm.

The imagistic construct of human cognition and culture change in View Four is more like a weather model than like a linear graph of human progress. That is, it presupposes that prior influences are still circulating and eddying about as new events introduce new patterns and possible "butterfly effects", much in the manner illustrated by Lorenz Equations in weather prediction studies or Strange Attractors used in models of other natural phenomena. An example of an

SIMULTANEOUSLY EXISTING WORLDVIEWS

Premodern
Mystical World
— Known through Ritual, Meditation, Spiritual Practice
— Purpose: Communion with Mystical World
Human as Spiritual Entity in Harmony w/Larger Cosmos/Diety/World Design
Cosmic Continuum of Time & Space/ Cyclical or Pulsing Time
Graphic Symbols: circle, spiral, wave

Modern
Rational Predictable Clockwork World
— Known through Observation and Logic
— Purpose: Description, Generalization, Prediction and Control
Mechanistic Human as Separate Discrete Entity, Generalization, Averages, Statistical Prediction Sought
Emphasis on Universality
Linear Sequential Time
Cartesian Grid of Time and Space (Myth of Progress)

Late Modern or Postmodern
Relativistic World
— Known through Observation of Breakdowns of Modern Paradigm
— Purpose: Liberation from Universalized World View and Concomitant Human Consequences
Human Unique, Alienated, No Generalization or Universality Possible Across Humanity
Fractured and Reassembled Grid of Time and Space
Electronic Pastiche

New Paradigm
Integral World:
— Complex, Non-Linear, Self Organizing, Self Regulating
— Known through Observation and Systemic Models & Simulations
— Purpose: Generation of Organic Process-Oriented View of Natural World with Human as Integral Part
Human Phenomena may be Modeled but Not Predicted
Simultaneous Unity and Diversity
— Studies in Cognitive Sciences, Biology, Weather & Origins of Universe Reveal Similar Forms
— Butterfly Effect: Small Variation Yields Large Effect
Time andSpace No Longer Separate (Unified Field)
Computer Graphics of Dynamic Complex Systems
— Simultaneously Regular and Unpredictable

FIGURE 12.1. Simultaneously Existing World Views

unexpected butterfly effect in this realm is the following: Boundary breaking changes from one world view to another often result from technologies designed to extend the existing view. For example, the premodern cyclical culture developed the technologies of time, calendar and clock to assist in ritual practices. These contributed to the linear time and grid based schemas prominent in modern culture. Modern culture developed the computer to assist in the generation of models based on logic and scientific observation. This is a View Two perspective. However, the computer is involved in pushing the boundaries of this view revealing blind spots, assumptions, and inadequacies.

The imagistic construct underlying View Two is the Cartesian grid. This illustrates an underlying realist assumption that a one to one correspondence of model with an objective reality is possible. This graphic symbol may be used to represent not only the design of conventional digital computing systems but time, space (mapped coordinates and pictorial representations), the format that underlies knowledge systems such as Library of Congress and Dewey Decimal systems, divisions of disciplines in universities, organizational staffing patterns in public and private institutions, the design of most textual prose. (Exceptions include the stream of consciousness fiction of James Joyce, William Burroughs cutups, and some contemporary interactive fiction. These are more closely related to View Three). The Cartesian grid is the basis for most work in computer graphics and vision systems. It contributes to the conceptual basis for the separation of art and science from the lived embodied experience of everyday life. See Figure One for a diagrammatic depiction of the use of the Cartesian Grid in common knowledge structures, social organization structures, etc.

In this paper culture change, cognition, and imagery are assumed to be interdependently related. That is, cognition shapes perception, understanding, and creation of imagery. The imagery created may contribute to culture change and consequent further changes in imagery and cognition. This transformative nature of imagery beyond communication (third section of the model presented) and the choice of View Four for the conceptual design of this paper are influenced by these assumptions.

12.2.1 New Research Paradigm (View Four): Implications for a New View of Variability and Universality

Recent research perspectives in the sociology of science may be best represented by the following assumptions:

1. Complexity is a genuine irreducible phenomenon. This was made evident via computer models of deterministic-recursive systems in which simple mathematical equation systems provide extremely complex behavior.
2. Irregularity of nature is normal, not an anomaly, and forms the focus of research. Non-equilibrium processes are recognized as the source of order and the search for equilibrium is replaced by search for dynamics of process.
3. Self Regulating Model of systemic closure replaces the classical system- environment model based on external control. Effects produced by the system are the causes of systemic organization and maintenance. In sufficiently complex systems internal self-observation and self-control form the basis of cognition. Any information a system provides on its environment is a system-internal construct. The "reference to the other" is merely a special case of "self reference" [7].

These assumptions were generated following the proliferation of computer graphic representations of complex systems in multiple research areas. They also follow upon the development of theory in biology based on autopoetic, self organizing systems. They illustrate a potential new view of interaction of imagery, information processing and consequently a change in mode of examining and understanding the world. They involve an epistemological shift in world view similar to those described in "Cultural Implications of Integrated Media" [6].

Varela, Thompson and Rosch [8] state:

> Largely because of cognitive science, philosophical discussion has shifted from concern with a priori representations (representations that might provide some noncontingent foundation for our knowledge of the world)

> to concern with a posteriori representations (representations whose contents are ultimately derived from causal interactions with the environment) (p. 137).

This shift could only occur as a result of technological developments in computer hardware and software.

12.2.2 Technological Developments and Understanding Imagery

Increasingly developments in computing hardware and software are shaping understanding of natural phenomena, including human imagery. Developments in three important areas have contributed to changes in understanding. These are measurement instrumentation, data analysis strategies, and hardware for interface and display. It is now possible to measure aspects of human behavior that were formerly invisible. For example, positron emission tomography permits generation of new images depicting brain processes that were formerly inaccessible. Data analysis techniques permit analyses of vast amounts of data in new ways, for example, clustering, factor analysis, multidimensional scaling, weighted decisions, and fuzzy logic strategies. Developments in interfaces, speed and memory allow researchers to interact in real time with three dimensional dynamic images of information, a cyberspace of data, in ways hitherto unimaginable.

The importance of contemporary interactive computer graphic imagery becomes evident in problems requiring depiction of simultaneous multiple relations and versions of a system or depictions attempting to make visual sense of chaotic systems and multidimensional structures. This approach is also important in the conceptual design of information systems encompassing complex dynamic information that require multiple individualized choices and routes through this information. The privileged position of alphanumeric representation for knowledge generation and manipulation is brought into question by the existence of new technologies of representation and information management. The grid has become insufficient as a structure in which to place knowledge. Visual electronic images have begun to take a prominent role in re-visioning knowledge construction in the arts, humanities and social sciences, as well as in the natural sciences. The boundaries between these disciplinary views of the world are becoming blurred by the transformability of digital information.

All of these factors contribute to changing views of human image processing.

An example of the effects of technological developments on cognitive science is the current tendency to approach cognition in a synthetic manner with the aim to relate a given cognitive function to its corresponding neural organization and activity state. Examples of this may be seen in the work of Aribib [9]. The work of Posner, Peterson, Fox, and Raichle [10] indicates that neuropsychology and neuro-imagery usefully complement the psychological approach by offering ways to dissect global functions into elementary operations that are localized in the brain. These may reframe the usual conceptual image of the brain and show that the cleavage of brain functions into the classical neurological, algorithmic, and semantic levels is no longer appropriate and may even be misleading ([11],pp. 98-99).

There is also current work in computer imagery depicting brain activity beyond the neuronal level to the receptor level and even to the molecular level. While this activity is based on premises rooted in View Two, i.e. realist assumptions involving intent to observe, model, and predict, it is raising interesting unexpected questions about the relationships of humans to one another and to their environments. The emergence of these questions serves as an illustration of the model presented in this paper.

An example of computer imagery that is related to the interactive nature of imagery, cognition, and culture change within the environment is the use of computer imagery for modeling molecular structures as a precondition for molecular engineering. Drexler [12] states:

> Although physical models cannot give a good description of how molecules bend and move, computer-based molecules can. Computer-based modeling is already playing a key role in molecular engineering.As John Walker (a founder and leader of Autodesk) has remarked, “Unlike all of the industrial revolutions that preceded it, molecular engineering requires, as an essential component , the ability to design, model, and simulate molecular structures using computers.” ...John Walkers remark was part of a talk on nanotechnology ... (p. 107).

Walker further speculates that “Current progress suggests the revolution may happen within this decade, perhaps starting within five years." (p. 116). This interrelationship of human cognition and technology present in computer graphic software and resulting imagery is an example of the second part of the model in this paper, communicating with self and others via imagery. Although most computer graphic imagery relies on the idea of one to one correspondence with objective reality, it also functions to stimulate new ways of thinking in the same manner as artistic image production.

12.2.3 Cultural Basis for Seeing Human Variability as Problematic

The new view of variability may be contrasted with the more common view attending a simpler scientific realist orientation to perception and understanding (View Two). View Two is focused on the norm and tends to disregard and devalue particularity, irregularity, and variability. A primary assumption of this view is that the world consists of predictable and manageable information. Historical roots of this view and reasons for its pervasiveness in the world of technologically created imagery are reflected in the influence of View Two on computer hardware and software, especially in computer graphics. My article, "Computer Graphics: Effects of Origins," discusses this [3]. In addition, View Two is an important influence on the quantitative psychological and sociological research that emphasizes the norm, standardization, predictability, and regularity of information.

Human variability due to individual and cultural differences is emphasized in Views Three. View Four and in the model presented in this paper emphasize simultaneous regularity and variability. This has been made possible because of advances in computer technology. Concomitantly, discussions of factors attending this perspective as it impacts creation of scientific visualization, graphic interfaces, electronic information, and communication systems and electronic art forms logically follow in a self-reflexive manner.

12.3 An Information Oriented Model for Understanding Imagery

This model focuses on the relation of human cognition and culture change stressing dynamic self-reflexive, self-organizing nonlinear patterns. It begins by examining the context for perception and understanding (individual, psycho-cultural climate within the immediate environment, visual physical environment both immediate and surrounding and the larger cultural context, i.e. social, cultural historical). Secondly, it examines observable processes that are involved (symbolic representation, delineation in material form, presentation, expression communication, evaluation, feedback, and transfer). Finally it addresses potential transformations of the original context resulting from these observable processes.

A MODEL OF HUMAN PERCEPTION, SYMBOLIC REPRESENTATION, MEANING DERIVATION, COMMUNICATION, AND TRANSFORMATION

I. FACTORS AFFECTING PERCEPTION AND DERIVATION OF MEANING

1. Individual World

Interior

1.1 Present physical condition

- 1.11 Genetic
- 1.12 Biochemical
- 1.13 Psycho physical
- 1.14 Neurological

Interface

1.2 Sensory Interface

- 1.21 Sight
- 1.22 Hearing
- 1.23 Smell
- 1.24 Taste
- 1.25 Kinesthetic, Spatial
- 1.26 Tactile
- 1.27 Awareness beneath conscious levels of each
- 1.28 Interrelationships among these, interior physiology and exterior environment

Interior, Interface, and Environmental Factors Affecting Perceptual, Conceptual, and Symbolic Development and Understanding within Individual

- 1.3 Cognitive style
- 1.4. Information Handling
  - 1.41 Adaptivity related to cognitive style
  - 1.42 Preferences for internal symbolic representation
    - 1.421 Visual (spatial, graphic, iconic, linguistic, mathematical)
    - 1.422 Auditory (tonal, linguistic)
    - 1.423 Kinesthetic (spatial, tactile)
  - 1.43 Prior experience and practice in information processing
    - 1.431 Amount in specific symbolic forms
    - 1.432 Types of structures used in processing information (influence of personal predilection, family dynamics, subcultural group, larger cultural context)
    - 1.433 Degree to which valuing of specific symbol systems match valuing of family, subcultural, and larger culture
  - 1.44 Values, attitudes, and beliefs affecting information handling patterns
    - 1.441 Values, attitudes, and beliefs which allow or restrict risk taking
    - 1.442 Degree of valuation of factors such as tradition, or in contrast, originality as characterized by fluency, flexibility, and originality
    - 1.443 Degree of recognition that information handling and knowledge structures are culturally and historically based
- 1.5 Prior personal experiences, for example learning relative to specific experience under consideration
- 1.6 Cultural effects on individual's perceptual conceptual learning

1.61 Values of individual's home, subculture, immediate and larger community

1.611 Aesthetic

1.612 Ethical/Moral

1.613 Social

1.614 Economic

1.615 Relations among values

1.616 Degree to which individual has adopted these valuing patterns

1.7 Personal values, attitudes, and beliefs affecting information handling patterns

1.71 Attitudes and beliefs about risk

1.72 Attribution of control, sense of self-worth, sense of personal agency and confidence, presence or absence of recent positive affect

1.73 Degree of valuation of factors such as tradition, or in contrast, originality as characterized by fluency, flexibility, and originality

1.74 Degree of recognition that information handling and knowledge structures governing information handling are culturally and historically based. (Example of grid and dynamic interactive, self-organizing computer graphic models as contrast between governing modern and contemporary knowledge structures)

1.8 Self perceived readiness for specific behavior under consideration

2. Psycho-Cultural Climate within the Immediate Context of Experience

2.1 Psychological differences among participants

2.2 Cultural differences among participants

2.3 Relative power of participants

2.4 Degree of fit among cultural valuing patterns of participants

2.5 Degree of fit among perceptual, cognitive, and affective structures of participants and their reactions to 3 below.

3. Visual-Physical Environment, Immediate and Surrounding

    3.1 Variety and degree of sensory stimuli present
        3.11 General environment
        3.12 Focused aspect of environment, for example a virtual environment
    3.2 Structure and design
    3.3 Physical condition of environment
    3.4 Resource availability (necessary tools, materials, space, and other resources)
    3.5 Meanings attributed to environment by participant based on factors in 1 and 2 above.
        3.51 Affective meanings attributed by participants (personal and shared)
        3.52 Symbolic meanings attributed by participants (personal and shared)
        3.53 Conscious interpretations of environment based on factors above.

4. Larger Context: Social, Cultural, and Historical

    4.1 Assumed Cosmology and Related Knowledge Structures
        4.11 Form and number of organizing systems present
        4.12 Effects upon forms of symbolic and material culture
            4.121 The real in graphic form (two and three dimensional)
            4.122 Iconic form (map example)
            4.123 Linguistic form
            4.124 Classification system formats (Indexing systems, organizational charts, taxonomies, outlines, etc.)
            4.125 Interplay among specific categorization patterns (Example color naming showing interplay among symbolic representations systems [linguistic and graphic] across cultures)
        4.13 Values, attitudes, and beliefs characteristic of primary cultural group over time

4.14 Relative influences of secondary or variant cultural groups at present

II. FACTORS AFFECTING DELINEATION, EXPRESSION, COMMUNICATION AND EVALUATION

5. Symbolic Representation, Delineation

5.1 Processes and materials used in formalizing understanding of topic under consideration within individual desiring to communicate

5.2 Translation of percepts and concepts into symbols and information patterns, which are believed to communicate information to other participants.

5.21 Choosing appropriate symbolic structures, to design and present information to intended segments of larger population.

5.22 Choosing appropriate media, tools, and presentation techniques, to match information in conceptual and affective intent to the intended audience

5.3 Delineation of material/symbolic representation which takes all the above factors into consideration

6. Presentation, Expression, Communication

6.1 Choice of context for presentation (Setting the Stage)

6.11 Audience selection, analysis, possible testing

6.111 Individual to self

6.1111 Thought representation only (Imagery Debate)

6.1112 Delineation that requires material/ symbolic representation

6.112 Individual to other(s)

6.113 Self to one other

6.114 Self to small group (Close or distant relationship)

6.115 Self to larger group

6.116 Group to group

- 6.12 Design of psycho-cultural environment
- 6.13 Design of immediate visual physical environment
- 6.14 Design of virtual communication environment
- 6.15 Determine and design specific links to larger context

6.2 Delivery choices

- 6.21 Presenter and participants
  - 6.211 Relationships of presenter and participants

(Virtual presence of creator of interface within use: Examples- 1. Musical interface lecture of Williams (this conference). 2. Kidpix interface. Both stimulate comfort and pleasure in user. Speculation regarding neurochemistry of pleasure as stimulating associativity and creativity.)

  - 6.212 Degree of participation by various participants
  - 6.213 Interaction patterns
  - 6.214 Power dynamics of human participants
- 6.22 Representation
  - 6.221 Thought Representation
  - 6.222 Symbolic Material Representation
  - 6.223 Choices of Symbolic Representation(s)
  - 6.224 Choices of Material Representation (s) (Media)
- 6.23 System
  - 6.231 Structure and Design of Delivery System
  - 6.232 Aesthetic dimensions of system
  - 6.233 Social and political dimensions of system
  - 6.2331 Power dynamics of system (examples in Jones (1991) integrated media article)
  - 6.234 Cultural influences of knowledge structures on system
- 6.24 Setting
  - 6.241 Psychological and cultural aspects of setting
  - 6.242 Visual and physical aspects of setting
  - 6.243 Characteristics of virtual communication setting

7. Evaluation, Feedback, and Transfer

- 7.1 Choice of evaluation perspective
  - 7.12 Choice of validity type(s)
  - 7.13 Choice of method(s) of evaluation
    - 7.131 Participant identification
    - 7.132 Procedures
    - 7.133 Instrumentation
- 7.2 Choice of object(s) of evaluation
  - 7.21 Presenter(s)
    - 7.211 Evaluation of motivation
  - 7.22 Participants(s)
    - 7.221 Evaluation of change in participants, individually and in context
    - 7.222 Evaluation of participants' roles in information handling
    - 7.223 Evaluation of motivation
  - 7.23 Information
    - 7.231 Evaluation of immediate and larger contexts relative to individuals and information transfer
  - 7.24 Presentation
    - 7.241 Evaluation of symbolizing, designing, and presentation of information
    - 7.242 Evaluation of choice of media, tools, and final product
    - 7.243 Evaluation of effectiveness in terms of audience
- 7.3 Application and purpose of Evaluation
  - 7.31 Audience
    - 7.311 Presenter
    - 7.312 Participants
    - 7.313 Larger Audience
  - 7.32 Intended purpose
  - 7.33 Degree to which purpose is met
  - 7.34 Feedback

7.4 Potential transfer (from one context to another)

7.41 Information

7.411 By Participants

7.412 By Presenter

7.413 By Evaluators or larger audience

7.42 Presentation 7.43 Evaluation

III. FACTORS AFFECTING SYMBOLIC TRANSFORMATIONAL ACTIVITY: CHANGING CULTURE AND COGNITION

8. Beyond Communication: toward Symbolic Transformational Activity

Activity

8.1 Individual

8.11 Self transformation via visualization Psychological (cognitive, i.e. changes in understanding or problem solving or affective or mental health) Physical (physical health and well being, sports rehearsal)

8.12 Self transformation via drawing or construction of images, pictures, diagrams, models, etc.

8.2 Immediate Inter-Individual

Impact

8.3 Immediate Visual Physical Environment

8.4 Larger Cultural Context Wider Audience, Setting and Time Frame Physical construction of reality Nanotechnology Example Imagistic constructs that underlie conceptual patterning of culture or age Cartesian Grid Example

Interaction

8.5 Self-Organizing, Self-Regulating Complexity as New Imagistic Construct and as Organizing Pattern for Human Image Processing

Although the model is presented in outline format, it should be viewed as a dynamic pattern with all parts interacting in complex patterns. Examples have been chosen to illustrate this nonlinear interaction. The first example addresses part four of the first section of the model. It is concerned with the larger social cultural and historical context. This aspect is often ignored in discussions of technological imagery, but is commonly addressed in anthropology and art. For example, Grigg states[13]:

> Travelers and anthropologists have often reported with surprise that peoples from other cultures do not feel at ease with the images we regard as natural and realistic. We have assumed that the technological superiority of our images would be widely acknowledged. Yet to peoples from other cultures our images appear unnatural and distorted....Their own images, which strike us as willfully conventional and stylized, seem realistic and lifelike to them. Today these reports concerning primitive tribesmen are commonplace. No one challenges them. Similar reports made by Chinese and Japanese informants in this and the last century have also gone unchallenged. The evidence in fact is overwhelming that individuals immersed in these cultures do not agree with us about what counts as a realistic or truthful image (p.399).

Not only other cultures but other times within a single culture may use different conventions to depict reality. For example, consider another time period within European culture. If Byzantine art is viewed by contemporary viewers they usually agree that it is not realistic. However, C. Mango [14] cites many instances that show that for almost a thousand years, various representatives of Byzantine culture have indicated that their own art is lifelike, writing for example, "The work of painters is constantly praised for being lifelike: images are all but devoid of breath, they are suffused with natural color, they are on the point of opening their lips in speech" ([14], p. 65).

It is interesting to examine images from other cultures and to compare them to contemporary technological imagery. For example, Australian Aboriginal bark paintings reflect a view of the world seen as webs, networks, and fields of energy. According to Lawler [15], before and during the physical appearance of an animal species, the

Aborigines imagine an energetic field, which is regarded as the spirit or dreaming of that species. Images depicting this had been painted to explain this perspective to European viewers. Most Europeans failed to understand. Some artists added more figurative elements to their work to assist the explanation. When Euro-Americans are presented with Aboriginal bark paintings depicting the universe as energy fields they do not generally read them as realistic representations. However, computer graphic simulations and visualizations used in contemporary physics and neuroscience that depict the universe or the human brain in terms of energy fields may be read by them as realistic. Further relationships of image to technological development and cultural context is clearly revealed in the following conversation:

> During a conversation with an Aboriginal elder I came to see clearly the effect of the loss of hunting and gathering on the deep workings of the human psyche. The old man explained that in the trance vision one can see a web of intersecting threads on which the scenes of the tangible world as well as dreams and visions are hung. "Inner fears," he said, "break that glimpse of an invisible web work leaving only a world of isolated things. Some of the young Aboriginal men today talk and act very smart, they no longer have the vision cause they have the same fear inside as the white fellas. That's because they cannot go off in the bush and feed themselves like I do. Anyone who does not know how to feed himself is always frightened inside like a little child who has lost his mother and with that fear the vision of the spirit world departs." Conversations with Bobby McLoed and Uncle Bul 1988 cited by Lawler ([15], p.373).

Oscar Wilde states that art is man's gift to nature. This refers to the ways in which art may change our perception of the world. Rollins [16]states that:

> It has, indeed been shown experimentally that canonical form operates in perceptual processes, for example, in the fact that outline drawings provide information more quickly than photographs, and it can be argued that any

> picture can embody such form[17]. Since canonical form provides a mental schema with the character of a prototype, which can be used in perception in general, there is a sense in which seeing pictures can in turn affect perception. An artist is able to combine within a single view features that could not normally be seen together in order to permit optimal representation of form for perception and also to provide canons of execution that are unambiguous. Thus, departures from central perspective and from strict projective fidelity are utilized in the interest of representation (p.101).

There is little doubt that the larger context affects individual perceptions of what is realistic. It also affects attention, perception and imagination. The interaction of the larger context and seemingly simple elements of individual perception and cognition are illustrated in the following brief examples treating attention, color perception, and sequential successive processing of information or parallel distributed processing of information. These three examples illustrate the layered interactive nature of the model. It illustrates the view inherent in the model that the human is integral with surroundings, not separated by a layer of skin, but joined to the world by skin and other sensory organs and exchange systems. This coincides with the systemic view inherent in the new paradigm.

12.3.1 Attention, Selection, and Interpretation

In the 1890's, William James evinced interest in the human capacity to focus attention on selected aspects of environment to the exclusion of others. He held that human attention is a key element in deciding what features of the phenomenal flux would be figure and what would be ground. Early figureground experiments in perceptual psychology show a continuation of interest in this topic. Contemporary studies of human visual perception indicate that very early in the process of perception/understanding some sorting of figure from ground, i.e. giving attention to selected aspects of the visual phenomenal flux, occurs.

A goal of early computer science research in human visual information processing was the simulation of human perception and interpretation of visual imagery. Initially, many studies focused on pattern recognition. Some of these studies relied upon simulating eye pattern movements for purposes of detecting edges and other features allowing separation of objects from background. Even this seemingly simple task is quite difficult. Results of early edge detection research is present today in inexpensive auto-trace features in microcomputer software. The inadequacy of these is evident if complex photographic material with uneven lighting is presented for auto tracing. The need for human attention, selection and interpretation in achieving a meaningful image is evident even in such simple instances.

Bela Julesz is a provocative and prolific researcher in visual information processing active since thc 1960s. Regarding attention and eye movements, Julesz notes that Helmholtz (in the 1896 third German edition translated by Nakayama and Mackeben 1989, not the first English edition translation used by most researchers)[18] was aware that serial shifts of focal attention underlying scrutiny can take place without eye movements. This was also extensively studied by Posner [19].

While not focusing specifically on the pattern recognition or edge detection research, Julesz further notes regarding measurement, focal attention:

> Unfortunately, while eye movements can now be measured with a 1 minarc accuracy (e.g. using the fourth-Purkinje image eyetracker of SRI), the position of focal attention cannot be measured yet. At present the time constant of PET-scan [positive emission tomography] is 1000 times slower than the rate of focal attention, and it is not yet conclusively proven that attention will increase the oxygen flow of those neural tissues that are engaged. Before some fast, non-invasive methods are found we will have no idea what the observer is looking at, except by asking him. Therefore, any scheme that wants to exploit an observers scan path measuring eye-movements is doomed to failure ([18],p.212-213).

He further states that this is analogous to a debate regarding depth perception between Brewster and Dove shortly after the invention of the stereoscope in 1838. A century after the debate Julesz and Fender [18] showed that the alignment necessary for stereopsis could be extended twenty times greater than originally believed. "Again what is the use of measuring observers convergence movements when a vast arena in depth is scrutinized by binocular-disparity tuned neurons." (p.213)

Bela Julesz writes about his work for psychologists, physicists, neurophysiologists, and philosophers interested in visual perception. He states:

> Recent advances in understanding focal attention have a great impact on connectionist neural networks. In these networks each neuron participates in learning and storing some information by a small amount as a result of changing the weights of connectivities to like neurons. This learning is rather indirect, and memories are stored as the global interaction of slow changes in the interconnection of the participating neurons....(such networks) are very different from the action of the human brain. In addition to focal attention wandering over a visual scene, attention can also scrutinize many processes within the human brain. While the reader can follow my thoughts, his attention can easily wander to many other thoughts in rapid succession, and return to this article. It is this rapid change of thoughts and mental processes that is so alien to algorithms and machines. (p. 213)

Attention and its role in separating the meaningful from background is key in the central theme of cognitive scientists Lakoff and Johnsons experiential cognition. This is encapsulated in the following citation:

> Meaningful conceptual structures arise from two sources: (1) from the structured nature of bodily and social experience and (2) from our innate capacity to imaginatively project from certain well-structured aspects of bodily and interactional experience to abstract conceptual structures. Rational thought is the application of very general cognitive processes— focusing, scanning, superimposition, figure-ground reversal, etc. to such structures [20].

Current researchers in connectionist neural net simulations are beginning to attend to factors such as attention, goal direction, emotion, and motivation [21]. Their work reflects a valuing of emotion as much as cognition. Characteristics traditionally valued by artists such as emotional expression, associative and alogical connections of information and reliance upon sensual perception are coming to be valued in multiple research fields. Some researchers in neurochemistry are working on the relationship of neurochemicals to factors such as attention and associativity in relation to positive affect, anxiety, and task performance. The latter relationships have also been researched in social psychology by Isen and her colleagues [22, 23, 24, 25], as well as by researchers interested in educational psychology and creativity. Linking positive affect with associativity, creativity and improved quality of performance has important implications for software design as well as many other applied fields.

Changeux and Dehaene cite the following recent research ([11], p. 98):

> The contribution of attention to the processing of sensory information is currently being investigated in great detail by joint psychological and neurological approaches [10, 26]. As in the case of the selection of meaning..., selections of actions and intentions may take place via a Darwinian mechanism among internally evoked and context-dependent pre-representations including complex chains of objects from the long-term memory stores. Combinatorial processes may produce novel intentions or inventions at this level. The selection will then be carried out by testing their realism from the cognitive as well as affective point of view. The connections existing between the limbic system and the prefrontal cortex offer a material basis for relationships between the emotional and cognitive spheres[27, 28, 29].

Researchers in the arts and human sciences are pursuing the role of attention in complex aspects of human perception/interpretation. The roles of attention and selection in human understanding, imagery creation and perception in the larger cultural context is present in the following examples of graduate research from the doctoral program I chair. These provide practical examples of the role of attention in relation to personal and cultural frameworks of values, attitudes, and beliefs. They also illustrate some of the difficulties of understanding imagery in complex natural settings. Rayala [30] discussed drawing as a tool to assist in the task of ethnographic observational research. Trained as an artist, Rayala was adept at observational drawing techniques in the Western cultural tradition of literal realism. In an initial pilot study prior to his dissertation he drew various aspects of the fishing boats of the subcultural group he was studying. His cultural informants would point out what he "did wrong" in the drawings, that is what he missed as crucial information that they could see as full participants in the culture. That is, they would respond to his drawing by identifying information that was important enough to them to be strongly differentiated, but conceptually invisible to him until they pointed it out. Once aware of the information, he was able to correct his drawing to reflect this change in understanding. He was able to attend to new information in the scene before him in this way. Studies such as his reveal the continual struggle needed to communicate and point to the role of attention in image perception. How do we provide the information that others need in order to see as we see, know as we know? How do we access their world of information so we can observe the differences between our world and theirs? The second part of the model in this paper treats these issues. Other studies that have used image creation and analysis to address these questions include Hoffman's study [31] of a historical piece material culture. Her study incorporated drawing and reconstruction of artifacts as means of study. Wasson's [32] and Yuan's [33] studies of aesthetic valuing in other cultural settings (Australian aboriginal sites and a Taiwanese temple site) utilized photographic and diagrammatic data, combined with interviews about what participants selected for attention, their attitudes, preferences and beliefs. This disparate data was analyzed using visually weighted iconic indexing and coding strategies, and sorting strategies that range from Boolean and weighted decision processes to intuitive and random. The purpose of these analyses was to generate hypotheses regarding what is worthy of study in the situation based on grounded data rather than a top-down analytic strategy. A presupposition of these studies is that time spent in grounded-category construction, interpretation, and hypotheses generation is a necessary precondition to grounded-theory generation. A series of studies using these techniques has been done under my direction. A recent study using these is Keifer-Boyds [34] analyses of human movement. It uses digital video, superimposed vector graphics, editing, and hypermedia techniques to assist in information analysis and presentation of findings. All of these studies are concerned with using visual analytic techniques for the purpose of analyzing disparate information. All stress the role of attention and selectivity of observers and researchers as integral to the research process.

12.3.2 Color

Cognitive science has begun to address the interaction of culture and perception. For example, how does the way a particular culture categorize phenomena affect perception of participants in a culture. As one example of the difficulties surrounding categorization tasks relative to visual imagery consider the case of research in color naming. Research in this area illustrates the difficulties in dividing perception and cognition, cultural and personal, linguistic and graphic, and other similar culturally influenced categorical divisions.

12.3.2.1 Example of Old and New Research Perspectives:

Early researchers in color perception adopted a View Two structure of thought. That is, they averred that color is a physically observable phenomenon that can be measured in terms of wave lengths of light with names attached to various sections of the spectrum. They expected that all humans would perceive and interpret these wave lengths similarly as a universal condition. The belief that complex phenomena such as visual information processing can best be studied by breaking it into reductionistic components such as color (wave lengths of light) and studying each component also contributed to the style of early studies. According to Varela et al. [8]:

> Recent computational models of color vision seem to support this line of argument. The surface reflectances of objects in our surrounding world, such as bricks, grass, buildings, etc., can be expressed in a rather limited (three dimensional set of prototypical functions[35, 36, 37]. Thus it would seem that all the visual system has to do is sample the scene with its three color channels and thereby reconstitute the surface reflectances from the activities in these channels. On the basis of these models several

> vision scientists, as well as certain philosophers, have argued not only that the function of color vision is the recovery of surface reflectances but that color itself is just the property of surface reflectances [30, 38]. For a criticism of this view see E. Thompson's forthcoming book, *Color Vision: A Study in Cognitive Science and the Philosophy of Perception.*

However, anthropologists interested in studying perception in context discovered many ways of dividing, naming and classifying colors in other cultures. Based on these cultural studies of color naming, social scientists with beliefs about the primacy of language in shaping perception assumed that different groups of people saw color differently. That is, they moved toward the third structure of thought and assumed a relativistic stance. An example of this style of thinking (Sapir Whorf hypothesis) is exemplified in this quote from Gleason [40]:

> There is a continuous gradation of color from one end of the spectrum to the other. Yet an American describing it will list the hues as red, orange, yellow, green, blue, purple, or something of the kind. There is nothing inherent either in the spectrum or in the human perception of it which would compel its division in this way.

Early research cited by Varela et al that supports this relativistic (View Three) stance includes experiments that demonstrated memory for colors (a cognitive variable) was a function of color naming (a linguistic variable). Among these were Brown and Lenneberg [41], "A study in language and cognition;" Lantz and Steffire [42], "Language and cognition revisited;" Steffire, Castillo Vales, and Morely [43], "Language and cognition in Yucatan."

However some researchers have conducted studies in which cultural groups with different color categorization patterns are asked to engage in sorting and matching tasks. These studies do not support the relativistic view. That is, even though the participants may name and classify colors differently they sort and match color chips in a similar manner. Most frequently cited of these studies is the 1969 study by Berlin and Kay [44], "Basic Color Terms: Their Universality and Evolution." Their study over ninety languages determined eleven basic color categories encoded in any language, though not all languages encode all eleven. Subjects speaking various languages were given color chips and asked to specify the boundaries and best examples of the colors to which their basic terms refer. Although there was considerable variation on category boundaries, there was significant agreement on the best example or foci of the color categories. This style of research treating a seemingly simple aspect of perception is echoed by the chapter on texture by Bhushan and Rao in this volume. Another example of effects of culture and environment on simple perception is the greater susceptibility of individuals who live in carpentered right-angled environments to the Muller Lyer illusion. Cross cultural studies reveal the complexity of simple perceptual processes in natural settings.

Another set of interesting studies on color categorization was done by E. Rosch (then E. Heider) with the Dani of New Guinea (E. Heider [47], "Universals in color naming and memory;" E. Heider [48], "Linguistic Relativity;" E. Rosch [45], "On the internal structure of perceptual and semantic categories;" E. Heider and Olivier [49], "The structure of color space in naming and memory for two languages." Their language lacked virtually all color vocabulary.

> (She) found that (1) central members of basic color categories were perceptually more salient, could be learned more rapidly, and were more easily remembered in both short-term and long-term memory than were peripheral colors, even by speakers of Dani who do not have names for the central colors; (2) the structures of the color spaces derived from Dani and English color naming were very different but were quite similar for those derived from Dani and English color memory; and (3) when Dani were taught basic color categories, they found it quite easy to learn categories that were structured in the universal fashion (with central members as central but extremely difficult to learn categories that were structured in a deviant manner (with peripheral colors as central, where blue-green might be central and blues and greens peripheral) ([8], p. 169). She also found very similar effects were found in the development of color names in young children in our own culture. (E. Heider [50], Focal color areas and the development of color names) (Cited in [8]).

That is not to say, however, that colors have the same meanings across cultures. The perspective that some aspects may be cross culturally similar and some different more nearly matches a View Four perspective and that of the model presented in this paper. One of the major differences in human color perception appears to be degree of personal experience in differentiating colors. For example, graphic artists constantly involved in careful differentiation and selection of colors for printing can differentiate more colors in color-matching tests than can persons with less experience. The standardized colors used in graphic arts and interior designs have many more colors than those used in color experiments such as those described above. To persons with little experience matching and sorting colors, many of these are indistinguishable while experienced persons can readily differentiate these. This example of color- categorization research illustrates the interactive nature of the model presented in this paper. It also illustrates the complexity of a seemingly simple component of visual-information processing.

The complexity of mechanisms involved in color perception can be seen in early color experiments of Land. A large distributed neural network is involved in color perception. Recent brain research indicates that there are twenty different brain areas involved in vision in macaque monkeys versus only two in rats [51]. Artists, for example, Kandinsky, and more recently cognitive scientists, have further noted the interrelationships of color with other sensory modalities. Cognitive expectancies and memories influence many aspects of perception in addition to color perception as evidenced by the prior discussion of realism in visual imagery.

12.3.3 Serial/Parallel Processing

Another area of interest of early researchers was the differentiation between simultaneous and successive information processing. Neisser [52, 53] linked visual perception with the capacity for simultaneous or parallel processing of information. Kaufman [54] cites him as indicating that "The defining criterion for an operationally parallel processing system is not necessarily simultaneity, but first and foremost independence, i.e. an operation executed in the system will not depend upon another operation. Operations may be executed serially, one at a time, independent of each other. ...visual imagery constitutes a parallel processing system in both the spatial and operational sense." (pp 53-54.) These obviously are important properties of visual imagery, since extra cognitive operations will take more time in sequential system, but not necessarily in parallel ones. A nice demonstration of parallel processing is found in an early experiment by Neisser et. al. [52].

A more recent text edited by Mandl and Levin [55] also discusses information processing of text and pictures. This discussion has been continued by neuropsychologists and by computer scientists working toward connectionist architectures in computing machinery.

Aystos' [56] essay, "Neuropsychological aspects of simultaneous and successive cognitive processes" provides an interesting overview of research bridging the time between the early research cited above and more recent research. Aystos specifically cites the work of Luria [57, 58] and the work of Das, Kirby and Jarman [59, 60] as influences on his own work. The latter emphasizes the role of modes of processing, i.e. information may be treated either as a simultaneous event or as a series of events. Aystos' work focused on neuropsychological factors as predictors of modes of processing. His work provided empirical support for the position of Das et al. It also supports the position that more study of the interaction of processing components with task type is necessary. This has considerable implications for many practical realms, for example, education, interface design, scientific visualization, and design of hypermedia knowledge systems. It also matches the underlying premises of the model presented in this paper. This position is being supported in the research reviewed by Hanson and Burr [61] in that hidden units allow relationship to task type in neural nets.

These three examples have been presented to show the complex interrelationships of parts of the model in three examples: attention, color, and parallel/serial processing. The following section discusses selections within the model that illustrate interrelationships of imagery, technology and the perspectives of the model.

According to William James:

> Mental knives may be sharp, but they won't cut real wood... With real objects on the contrary, consequences always accrue; and thus the real objects get sifted from the mental ones, the things from our thoughts of the fanciful or true, and precipitated together as the stable part of the whole-experience-chaos, under the name of the

> physical world. Of this our perceptual experiences are the nucleus, they being the originally strong experiences. We add a lot of conceptual experiences to them, making these strong also in imagination, and building out the remoter parts of the physical world by their means; and around this core of reality the world of laxly connected fancies and mere rhapsodical objects floats like a bank of clouds. This implies that we actively participate in the creation of our perceptions via use of conceptions.

In the view of Finke [62] this would agree with his third category of theories on the relationship of imagery to perception. His categories include:

1. Structural theories proposing that mental images exhibit the same spatial and pictorial properties as real physical objects.
2. Functional theories proposing that the formation and transformation of mental images contribute to object recognition and comparison.
3. Interactive theories proposing that imagery contributes directly to the ongoing perceptual processes.

12.3.4 Early Research on Human Image Processing

Quite early, psychologists had been interested in differences in human information processing of visually perceived alphanumeric, iconic, or pictorial information. Other researchers were interested in individual differences in self-reported preference for internal symbolic representations. In the 1960s and 1970s quite a number of psychological researchers investigated the relationship of visual information and verbal information to problem solving. Geir Kaufman [54] provides an interesting account of much of this research that continues to remain relevant. The following concepts of the researchers cited by Kaufman bear upon questions of imagery and high-level thought.

Berlyne [63] holds that the reason why visual imagery should be especially adapted to the representation of symbolic transformational responses is that visual imagery enables the subject to represent to himself both some actions of his own that may modify external stimuli, and such physical processes in the environment as bring about changes, independently of anything he might do. Words and combinations of words are, however, regarded by Berlyne to be poorly suited for the representation of transformations, because words can only effect changes in the environment through social mediation, i.e. communication. Arnheim [64] holds, in principle, a similar view, and regards visual imagery as the primary vehicle of productive thinking, while language is assigned a static and stabilizing function.

From a social-psychological orientation Sarbin [65] also points to the action-like properties of imagery in his view of creativity as an instance of muted role-taking, or 'as if' behavior carried out in mental imagery.

In an early work Pylyshyn [66] treats the issue of the place of imagery in cognition from an information-processing point of view, and arrives at a similar notion of the role of imagery in transformational activity.

Pylyshyn argues that images may be especially useful in the process of constructing new information: "...while picture-like entities are not stored in memory, they can be constructed during processing, used for making new interpretations (i.e. propositional representations), and then discarded" ([66], p.19).

Such a visual model, according to Pylyshyn, functions like a selective, abstract and interpreted percept, which has a special transformational role: "Its importance arises from the fact that it makes possible certain kinds of restructuring and reconstruction of descriptions" (p.19).

On this last point, Pylyshyn seems to build largely on an account of visual imagery given by Chase and Clark [67], who studied mental operations in the comparison of sentences and pictures. As to the function of images, Chase and Clark state that 'they are used to devise new structures. The mind's eye is a working space, if you will, for generating new concepts and relations' ([67], p.226).

Chase and Clark refer to chess experiments which indicated that abstract relations are constructed from more primitive relations:

> Since a complete retinal representation of the board is impossible, the player's search through the problem space is assumed to take place in visual imagery ...the subjects are in fact abstracting new information form the visual image in much the same way as they would from a physical stimulus. The surface transformations of the image are

> probably analogous to actual physical transformations of a perceived object (p.229).

This is held by Chase & Clark to be a phenomenon of considerable generality: "We might further suppose that these visual-imaginal processes are quite common in other problem-solving tasks, which require spatio-mathematical reasoning. If this speculation is correct, then the mind's eye serves an indispensable role in cognitive functioning" ([67], p.229). The general conclusion reached by Chase & Clark is that comparison operations are conducted at the abstract level, while generation of information through surface transformations occurs in visual imagery.

The views presented above seem to converge in the hypothesis that visual imagery may be especially adapted to the representation of transformational activity needed in tasks possessing a high degree of novelty. This may be due to visual imagery lending itself easily to the simulation of action needed for the generation of new and relevant knowledge ([54], p. 58-59). This is particularly relevant to contemporary work of computer scientists working in scientific visualization, simulation, and modeling. It is also relevant to the last section of this paper dealing with the transformational qualities of imagery beyond communication.

12.3.5 Recent Research in Human Image Processing

Recently neuroscientists have begun to study effects of internal imagery, sensory stimulation, and human interpretation on the biochemical and electrical behavior of the neuronal systems. The effects of these neuronal changes on cognitive and affective behaviors, i.e., memory, learning, and other responses, have tremendous implications for researchers wishing to understand human understanding of imagery. Art historians examining the power of imagery, medical practitioners and sports trainers using visualizations to facilitate change in their clients, designers of computer interfaces and scientific visualization procedures seeking to understand the relationship of display to problem solving and inventive thinking, educators using multimedia environments for learning and other practical researchers may be seen as having overlapping concerns when viewed in light of this research.

12.3.6 Imagery Debate

Major positions in the imagery debate taken by three computer scientists employ the simple cognitivist paradigm and Western conventions of pictorial realism. Kosslyn's early work implies that images are not language-like symbolic representations, but bear a nonarbitrary correspondence to the thing being represented. Kosslyn has formulated a model by which images are generated in the mind by the same rules that generate images in computer displays: the interaction of language-like operations and picture-like operations together generate the internal eye. Pylyshyn, considered a hard- line cognitivist by some, holds that the mental scanning and mental rotation of images are cognitively penetrable by the subjects beliefs, goals, and tacit knowledge. Thought is propositional and imagery is a special case of proposition-based thought. Fodor suggests that image types may vary; that is, images convey some information discursively and some information pictorially.

In contrast, Rollins [16] takes a deconstructionist position. He develops a theory linking studies of imagery with work on perceptual categories and picture perception and integrates these categories with the psychology of belief and desire. His work emphasizes the importance of nonlinguistic representations. He is attempting to lay the foundation for a theory of nonpropositional representation. His work is in contrast with the prevailing image debate positions of pictorialism and descriptionism.

He argues for the importance of "aesthetic psychology," requiring that biological and environmental factors be placed within the bound of consideration in understanding imagery. He is concerned with context and the emotional content of imagery:

> Consider, first Proust's observation that when studying faces, we measure them, but as painters, not as surveyors. Thus at one time, a face may appear thin and gray, at another, smooth and glassy, at a third, waxy and fluid. These are arguably perceptible properties, and the point they illustrate is this: While we can simultaneously **believe** that each set applies to the same face, we cannot **see** them in it at one and the same time. Thus seeing the face as having them is to adopt an attitude, the content of which determines to a certain extent the logic of beliefs and desires that ensues (p. 97).

This position is more congruent with the model presented in this paper than simple cognitivist positions. It also corresponds with research on figureground images and implies that human behavior can be expressed in non propositional attitudes. It is in sharp contrast with most computability-of-mind theorists as well as contemporary eliminative materialists such as Churchland.

Varela, Thompson and Rosch [8] imply the problems with cognitivism arose from too great a departure from a biological perspective. Two widely acknowledged deficiencies of cognitivism are presented:

> ...symbolic information processing is based on sequential rules applied one at a time. This von Neumann bottleneck is a dramatic limitation when the task at hand requires large numbers of sequential operations (such as image analysis or weather forecasting), a continued search for parallel processing algorithms has met with little success because the entire computational orthodoxy seems to run precisely counter to it....
>
> ...symbolic processing is localized: the loss or malfunction of any part of the symbols or rules of the system results in a serious malfunction. In contrast, a distributed operation is highly desirable, so that there is at least a relative equipotentiality and immunity to mutilations (p.86).

Rolls [51] discusses visual information at the neuronal level. He provides evidence for ensemble encoding, distributed representation in various parts of the visual system. For example, the amygdala is involved in making associations between visual representations of objects and primary reinforcing stimuli leading to pattern association neuronal networks. The hippocampus is involved in recognition and episodic memory using auto-associational neuronal networks. The advantages of distributed representations for the inputs to such associational neuronal networks include pattern completion, generalization, and graceful degradation (or graceful performance with an imperfect specification of the connectivity during development).

Rolls begins to give us some clues regarding the problem of visual constancy, an issue that has been perplexing to many disciplines:

> The finding that some neurons in this region of temporal visual cortex have object-based rather than view-

> centered responses provides a form of invariance which is very useful in providing an input to associative memories in for example the hippocampus and the amygdala (see Rolls [49]) for it means that the associative memories can store a single rather than multiple representations of the object, and moreover can respond appropriately when a different view of an object is seen (p.344).
>
> In order to form new (declarative) memories, including memories of visual objects such as faces, limbic structures such as the hippocampal system and perhaps the amyglda are necessary...Although these limbic structures are necessary for the formation of these memories, the memories are not stored in the limbic structures in the long term but instead are stored in the neocortical areas which perform, for example, visual processing, in that old memories are at least partially spared in patients with damage to limbic structures.

Psychologists, sociologists, anthropologists, art educators, and other educational researchers have studied other individual and cultural differences in perception, representational strategies, and communication conventions. They have studied effects of internal imagery, external imagery below the level of conscious awareness, and differences in effects of still and moving imagery, differences in effects of original art works and reproductions and a host of other topics related to how humans process and respond to internally generated and externally presented visual imagery. Art historians, art educators, and anthropologists have studied cultural and temporal differences in the reality constructs underlying visual imagery, as well as differences in the way images are regarded, used, and presented. For example, the power of the visual image has been regarded quite differently at different times and at different periods. In contemporary culture visual imagery is commanding a more and more important role in popular culture, communication technologies, and as a means of modeling, analysis, and presentation in the physical sciences. Sociologists and political scientists are examining the treatment of images in media critically. A paper presented by Bajuk at the conference from which this volume originated, critically examines modeling and visualization within technological image construction. These discussions of the effect of visual and aesthetic conventions on the information bearing qualities of imagery echo the earlier concerns of anthropologists using film and video. Their questions addressed relation of image to reality, selection of images, as well as aesthetic and scientific conventions governing image generation and presentation. However, a preponderance of academics within many disciplines continue to utilize conventional written text as the primary and most validated form of scholarly expression without a similar examination. In addition, many continue to use the conventions of only one discipline on their chosen symbolic representation rather than examining the potential of perspectives from other disciplines for contributions to their research. Both of these convention-based practices have their roots in a View Two world view based in conventions of linear text and divisions of disciplines.

This paper presents a model based on a variety of disciplinary perspectives related to imagery. Speculations are presented regarding implications of the new research paradigm for the development of computer graphic applications as well as other practical applications. The role of associative thinking and other aspects of "artistic" thinking styles for computer scientists and engineers are considered. The significance of human science and humanities research are also considered.

Art educators, because of their practical interests in human understanding of images have for many years attempted to understand and synthesize the research of other disciplines. Primarily they have studied the work of psychological, and social science researchers that treat visual perception, image generation and interpretation. V. Lowenfeld in *Creative and Mental Growth* [68] introduced psychological research into art education. J. McFee emphasized both psychological and cultural research in her work. She developed a model of human perception and delineation in her dissertation and later in *Preparation for Art* [69]. This model focused on perception and creation of art and was intended to assist art educators in integrating psychological and cultural research into their classroom practice. This model served as an important originating source in the development of the model presented in this paper. I have attempted to extend the work of these researchers by including information from past and recent multidisciplinary research.

Computer graphic imagery was being developed simultaneously with the research described in this paper. Early development was primarily for practical purposes and sponsored by government and industry. Scientists were beginning to realize and use the power of images for understanding complex data. Scientific visualization and simulation laboratories reliant upon computer graphic imagery were created. Others realized the efficiency and power of computer graphic interfaces. Complex information systems began to utilize graphics to assist users in navigating information. However, these and similar practical developments were often independent of the research in human image processing cited above. Only relatively recently have considerations of the psychological or socio-cultural human dimension of these phenomena played an increasingly important role in these practical developments. A recent thesis by R. Bockelman [70], a software engineer, reflects this. It is entitled, *Design Concepts for Internationalization of a Graphic User Interface.*

It proposes an object oriented programming model for construction of a computer graphic interface intended to accommodate selected cultural similarities and differences of computer users in various parts of the world. Such a model provides an initial way to consider constructing interfaces that accommodate individual differences as well as cultural differences. This style of thinking about interface construction could lead to addressing issues such as those raised by the research regarding modes of information processing, relationship of information processing to task type, psychological research in individual variability and universality and anthropological research in cultural variability and universality. Standardization versus customization of symbolic and material representations, products and services represent practical outcomes of contrasting conceptual treatment of software construction.

A higher level of conceptual integration may be reached if information across disciplines may be integrated into software development for artificial intelligence such as neural net architecture. Julesz [18], Varela [8], and Rolls [51] note a need for a new algorithmic approach to modeling cognition because of the characteristics of distributed, parallel and sequential processing. Carpenter, Grossberg, and Reynolds [71] describe a system that features efficient recognition for a non-stationary environment featuring many-to-one and one-to-many learning. This parallels Rolls discussion of visual constancy and distributed processing in biological neural networks. Hanson and Burr [61] describe explorations of an alternative to the rule based approach of simple cognitivism using neural nets. They answer the criticisms of Fodor and Pylyshyn [72] that connectionism merely provides an alternative hardware for implementing symbolic models with the latter's characteristic properties of "productivity", "systematicitICy", etc.. Hanson and Burr focus on unique properties of nets (subsymbolic or distributed representations)[73, 74], but although hidden units have special representational properties these do not amount to a new kind of representational approach. Instead, these properties turn out to be related to existing data reduction techniques in multivariate statistics and psychometrics [75, 76]. Hanson and Burr indicate that neural networks with recurrence can be described by a set of nonlinear equations; the solutions of such a system of equations are usually called the attractors or stationary points of the system. Of particular interest is the use of "hidden units". They also note that there is a close analogy between multidimensional scaling methods and nets with hidden units. They note that *An enormous increase in computational power seems to result from the relatively simple step of adding hidden units.* This concurs with the perspective stated earlier that most nontrivial hardware and software problems are conceptual rather than technical problems. However, the conceptual change may be dependent upon prior technological developments. Conceptual advances, indeed research and scholarly life in general, are attempts to push the bus on which we are riding or, as Varela et al suggest, lay down the path in the walking. Unpredictable consequences and butterfly effects may result from these activities that transform prior concepts of the real and the possible.

12.4 References

[1] Kaufman, A. Ed. (1990). *Proceedings of the First IEEE Conference on Visualization: Visualization.* San Francisco Oct. 23-26. Los Alamos California: IEEE Computer Society Press.

[2] Jones, B. J. (1989). Computer imagery: Imitation and representations of reality. *Leonardo* ,Supplemental Issue:Computer Art in Context, 31-38.

[3] Jones, B. J. (1990). Computer graphics: Effects of origins *Leonardo*, Supplemental Issue: Digital Image, Digital Cinema, 21-30.

[4] Jones, B. J. (1991). Cognitive sciences:Implications for art education. *Visual Arts Research*, 17 (1), 23-41.

[5] Jones,B. J. (1992). Cultural maintenance and change: Currents in art and technology Lecture, Third International Symposium of Electronic Arts, Sydney, Australia and article in *Media Information Australia* (forthcoming).

[6] Jones, B. J. (1991). Cultural implications of integrated media, *Leonardo*, Supplemental Issue: Connectivity Art and Interactive Telecommunications, 24 (2), 153-158.

[7] Krohn, W., Kuppers, G. and Nowotny, H. (1990). *Self Organization: Portrait of a Scientific Revolution.* Boston, Massachusetts: Kluwer Academic Publishers.

[8] Varela, F. J.; Thompson, E. and Rosch, E. (1993). *The Embodied Mind: Cognitive Science and Human Experience.* Cambridge: MIT Press.

[9] Arbib, M. A. (1985). *In Search of the Person.* Amherst, MA: University of Massachussetts Press.

[10] Posner, M. I., Peterson, S. E., Fox, P. T. and Raichle, M. E. (1988). Localizations of cognitive operations in the human brain. *Science*, 240, 1627-1631.

[11] Changeux,J.-P. and Dehaene, S. (1990). Neuronal models of cognitive functions. In Eimas, P. & Galaburda, A. M.(Eds) *Neurobiology of Cognition.* Cambridge: MIT Press, 63-109.

[12] Drexler, K. E. and Peterson, C. with Pergamit, G. (1993). *Unbounding the Future: The Nanotechnology Revolution.* New York: Quill William Morrow.

[13] Grigg, K. (1984). Relativism and pictorial realism, *Journal of Aesthetics and Art Criticism*, 42 (4), 397-408.

[14] Mango, C. ed. and trans. (1972). *The Art of the Byzantine Empire.* Englewood Cliffs, xiv-xv. See also his "Antique Statuary and the Byzantine Beholder." Dumbarton Oaks Papers, 17, (1963). 65ff.(Cited by Grigg)

[15] Lawlor, R. (1991). *Voices of the First Day: Awakening in the Aboriginal Dreamtime.* Rochester Vermont: Inner Traditions International, Ltd.

[16] Rollins, Mark (1989). *Mental Imagery: On the Limits of Cognitive Science.* New Haven: Yale University Press.

[17] Hochberg, J. (1978). *Perception.* 2nd edition. Englewood Cliffs: Prentice Hall.

[18] Julesz, B. (1991). Essay on early vision, focal attention and neural nets,pp. 209-216. In *Neural networks: Theory and Applications* (pp. 209-216). Richard J. Mammone and Yehoshua Zeevi (Eds.) Boston: Academic Press, Inc; Haracourt Brace Jovanovich, Publishers.

[19] Posner, M. I. (1980). Orienting of attention. *Quarterly Journal of Experimental Psychology,* 32, 3-25.

[20] Lackoff,G. (1988). *Cognitive Semantics in Meaning and Representation,* Umberto Eco et al (ed.) Bloomington: Indiana University Press.

[21] Levine, D. S. and Leven S. J.(1992). *Motivation, Emotion and Goal Direction in Neural Networks.* Hillsdale, New Jersey: Lawrence Erlbaum Associates.

[22] Isen, A. M. and Daubman, K. A. (1984). The influence of affect on categorization. *Journal of Experimental Psychology,* 47, 1206-1217.

[23] Isen, A. M. (1984). Toward understanding the role of affect in cognition. In R. Wyer and T. Srull (Eds.), *Handbook of social Cognition* (pp.174-236). Hillsdale, NJ: Erlbaum.

[24] Carnevale, P. J. D., and Isen, A. M. (1986). The influence of positive affect and visual access on the discovery of integrative solutions in bilateral negotiation. *Organizational Behavior and Human Decision Processes,* 37, 1-13.

[25] Isen, A.M., Daubman, K. A. and Nowicki, G.P. (1987). Positive Affect Facilitates Creative Problem Solving. *Journal of Personality and Social Psychology,* 52 (6), 1122-1131.

[26] Posner, M. I. and Presti, D. F. (1987). Selective attention and cognitive control. *Trends in Neuroscience*, 10, 13-17.

[27] Goldman-Rakic, P. (1987). Circuitry of the primate prefrontal cortex and the regulation of behavior by representational knowledge. In V. Mountcastle and K. F. Plum (Eds.), *The Nervous System: Higher Functions of the Brain. Volume 5, Handbook of Physiology.* Washington, D. C.:American Physiological Association.

[28] Nauta, W. J. H. (1971). The problem of the frontal lobe: a reinterpretation.*Journal of Psychiatric Research*, 8, 167-187.

[29] Nauta, W. J. H.(1973). Conections of the frontal lobe with the limbic system. In L. V. Laitiven and K. E. Livingston (Eds.), *Surgical Approaches in Psychiatry.* Baltimore, MD: University Pack Press.

[30] Rayala, M. (1983). *Pictorial Ethnography: A Descriptive Study of the History, Theory, and Practice of Drawing as a Researach Tool in the Social Sciences from 1800 to 1983.* Unpublished doctoral dissertation. University of Oregon. Eugene, Oregon.

[31] Hoffman, E. (1991). *The Murder Quilt: A Methodological Study Exploring Selected Research Methods, Techniques, and Procedures Used to Study Material Culture.* Unpublished dissertation. University of Oregon, Eugene, Oregon.

[32] Wasson, R. (1983). *The Construction of an Indiex for Hypothesis Generation Concerning the Art of Australian Aborigines in the Process of Culture Change.* Unpublished doctoral dissertation. University of Oregon, Eugene, Oregon.

[33] Yuan, J. (1986). *An Exploratory Study for the Purpose of Generating Hypotheses Concerning the Emic Aesthetic Valuing of a Group of Taiwanese Temple Participants Regarding the Temple Art.* Unpublished doctoral dissertation.University of Oregon, Eugene, Oregon.

[34] Keifer-Boyd, K. (1993). *An Exploratory Study of Nonverbal Digital Video Interactive Analytic Techniques Applied to an Individual Learning Dance.* Unpublished doctoral dissertation. University of Oregon. Eugene, Oregon.

[35] Maloney, L. T. (1985). *Computational Approaches to Color Constancy.* Technical Report 1985-01, Stanford University Applied Psychological Laboratory.

[36] Maloney, L. T. and Wandell, B. A. (1986). Color constancy: A method for recovering surface spectral reflectance. *Journal of the Optical Society of America.*

[37] Gershon, R. (1986). *The Use of Color in Computational Vision.* University of Toronto Technical Reports on Research in Biological and Computational Vision: RCB.

[38] Hilbert, D. R. (1987). *Color and Color Perception: A Study in Anthropocentric Realism.* Stanford Center for the Study of Language and Information.

[39] Thompson, E .(Forthcoming). *Color Vision: A Study in Cognitive Science and the Philosophy of Perception.*

[40] Gleason, H. A. (1961). *Introduction to Descriptive Linguistics.* New York: Holt, Rinehart and Winston.

[41] Brown, R. W. and Lenneberg, H. (1954). A study in language and cognition. *Journal of Abnormal and Social Psychology,* 49, 454-462.

[42] Lantz, D. and Steffire, V. (1964). Language and cognition revisited. *Journal of Abnormal and Social Psychology,* 69, 471-481.

[43] Steffire, V., Castillo Vales, V. and Morely, L. (1966). Language and cognition in Yucatan: A cross-cultural replication. *Journal of Personality and Social Psychology,* 4, 112-115.

[44] Berlin, B. and Kay, P. (1969). *Basic Color Terms: Their Universality and Evolution.* Berkeley: University of California Press.

[45] Rosch, E. (1973). On the internal structure of perceptual and semantic categories. In T. Moore (Ed.) *Cognitive Development and Acquisition of Language.* New York: Academic Press.

[46] Rosch, E. (1978). Principles of categorization. In E. Rosch and B. B. Lloyd (Eds.) *Cognition and Categorization.* Hillsdale, New Jersey: Lawrence Erlbaum.

[47] Heider, E. (1972). Universals in color naming and memory. *Journal of Experimental Psychology,* 93, 10-20.

[48] Heider, E. (1974). Linguistic relativity. In *Human Communication: Theoretical Explorations.* A. L. Silverstein (Ed.). New York: Halstead Press.

[49] Heider, E.and Olivier, D. C. (1972). The structure of color space in naming and memory for two languages. *Cognitive Psychology,* 3 ,337-354.

[50] Heider, E. (1971). Focal color areas and the development of color names. *Developmental Psychology,* 4, 447-455.

[51] Rolls, E. T. (1991). Information processing in the temporal lobe visual cortical areas of macaques. In Michael Arbib and Jorg-Peter Ewert, *Visual Structures and Integrated Functions.* New York: Springer-Verlag. 339-352.

[52] Neisser U. (1963). Decision time without reaction-time: Experiments in visual scanning. American *Journal of Psychology,* 76, 376-385.

[53] Neisser U. (1967). *Cognitive Psychology.* New York: Appleton-Century-Crofts.

[54] Kaufman, G. (1979). *Visual Imagery and Its Relation to Problem Solving: A Theoretical and Experimental Inquiry.* Irvington-on-Hudson, N. Y.: Columbia University Press.

[55] Mandl, H. and Levin, J. R. (1989). *Knowledge Acquisition From Text and Pictures.* Amsterdam, Netherlands: Elsevier Science Publishers.

[56] Aystos, S. (1988). Neuropsychological aspects of simultaneous and successive cognitive processes. In *Cognitive Approaches to Neuropsychology.* Williams, J.M. and Long, C.L. (Eds). New York: London Plenum Press, pp. 229-272.

[57] Luria, A. R. (1966a). *Human Brain and Psychological Processes.* New York: Harper and Row.

[58] Luria, A. R. (1966b). *Higher Cortical Functions in Man.* New York: Basic Books.

[59] Das, J. P., Kirby, J. R. and Jarman, R. F. (1975). Simultaneous and successive synthesis: An alternative model for cognitive abilities. *Psychological Bulletin*, 82, 87-103.

[60] Das, J. P., Kirby, J. R. and Jarman, R. F. (1979). *Simultaneous and Successive Synthesis: An Alternative Model for Cognitive Processses.* New York: Academic Press

[61] Hanson, S. J. and Burr, D. J. (1991). What connectionist models learn: Learning and representation in connectionist networks. In *Neural Networks: Theory and Applications.* Richard J. Mammone & Yehoshua Zeevi (Eds.) Boston: Academic Press, Inc; Haracourt Brace Jovanovich, Publishers, 169-208. Reprinted from (1990) *Brain and Behavioral Sciences,* 13 (3), 471-511.

[62] Finke, R. A. (1985). Theories relating mental imagery to perception. *Psychological Bulletin*, 98 (2), 236-259.

[63] Berlyne, D. (1965). *Structure and Direction in Thinking.* New York: Wiley

[64] Arnheim, R. (1969). *Visual Thinking.* Berkeley and Los Angeles: University of California Press.

[65] Sarbin, T. R. (1972). Imaging as muted role-taking: A historical-linguistic analysis. In P. W. Sheehan (Ed.), *The Function and Nature of Imagery.* New York: Academic Press.

[66] Pylyshyn, Z. W. (1973). What the mind's eye tells the mind's brain: A critique of mental imagery. *Psychological Bulletin*, 80 (1), 1-23.

[67] Chase, W. G. and Clark, H. H. (1972). Mental operations in the comparison of sentences and pictures in L. Gregg (Ed.) *Cognition in Learning and Memory.* New York: Wiley.

[68] Lowenfeld, V. (1947). *Creative and Mental Growth.* New York: Macmillan.

[69] McFee, J. (1961). *Preparation for Art.* Belmont, CA: Wadsworth.

[70] Bockelman, R. (1992). *Design Concepts for Internationalization of a Graphic User Interface.* Unpublished master's thesis, University of Oregon, Eugene, Oregon.

[71] Carpenter,G. A., Grossberg, S. Reynolds, J. H. (1991). A Self-Organizing ARTMAP Neural Architecture. In Richard J. Mammone and Yehoshua Zeevi (Eds.) *Neural Networks: Theory and Applications,* Boston: Academic Press, Inc; Haracourt Brace Jovanovich, Publishers, 43-80.

[72] Fodor , J. A. and Pylyshyn, Z. W. (1988). Connectionism and cognitive architecture: A critical analysis. *Cognition*, 28, 3-71.

[73] Rumelhart, D. E. & McClelland, J. J. (Eds.) (1986).*Parallel Distributed Processing: Explorations in the Microstructure of Cognition. Volume I:Foundations.* Cambridge MA: Bradford Books/MIT Press.

[74] Smolensky, P. (1988). On the proper treatment of connectionism, *Behavioral and Brain Sciences,* 11 (1), 1-59.

[75] Dunn-Rankin (1983).*Scaling Methods.* Hillsdale, NJ: Erlbaum.

[76] Everitt, B. (1975). *Cluster Analysis*, London: Heinemannn Educational Books.

13

Aesthetics and Nature

The Manufacturing of an Authoritative Voice in Scientific Visualization

Mark Bajuk[1]

13.1 Introduction

First off, I feel it is important to position myself so that you may understand the sort of work the ideas in this paper come from. I worked for four years at the National Center for Supercomputing Applications (NCSA) as a member of the Scientific Visualization Group. Within this group, I collaborated with academic researchers and corporate partners in the production of high-end scientific visualizations as well as educational graphics for popular scientific programs broadcast on US, Japanese, and Canadian television. This paper is a study of this particular visualization effort. Although my observations could, perhaps, be considered limited to a very particular subset of the visualization field, I believe that the issues discussed here are relevant to the wider practice, including interactive workstation-based systems and virtual reality.

This paper is divided into three sections. The first section investigates how decisions based on visual aesthetics and "good design" principles can manufacture a voice of authority for a scientific visualization. The second section analyzes the multiple social roles that the creators of scientific visualizations operate within, and how these different cultural situations influence the aesthetic decision-making process. The third section presents visualization case studies that illustrate some of the ideas discussed in the first two sections.

[1] 1631 Denniston Avenue, Pittsburgh, PA 15217

13.2 Scientific Visualization as a Coherent Window into a (Fragmented) World

Since its inception, advocates of scientific visualization have emphasized how it provides an audience with a unified, intuitive window into the inherently fragmented, discontinuous world created by a computer [1]. The ability of the human eye to quickly recognize changing subtleties in form and to create meaningful patterns out of complex arrangements of shapes is seen as a perceptual advantage to be exploited by the visualization process. There is a conscious effort on the part of many visualization producers, software developers, and researchers to present objects derived from simulations in a way that emphasizes an internal logic, order, and completeness. This occurs at the expense of emphasizing, for example, the spatial and temporal discontinuities of the data.

It appears that many of the design principles and preferences that go into the creation of a visually cohesive visualization lie fundamentally in a desire to imitate what we understand to be nature. I use the word "nature" here with the understanding that nature itself is a social construct, continually transforming and undergoing reinscription of meaning [2]. Visualization operates as an active participant within this culturally defined realm of nature, and functions as a significant contributor to the bridging between what is commonly understood to be "real nature" (e.g., an idyllic, undisturbed wood, water molecules, the solar system) and "artificial nature" (e.g., biological laboratories, national parks, computer simulations, museum dioramas).

In order to minimize the conceptual distance between "real nature" and "artificial nature," creators of visualizations manufacture or emphasize temporal and spatial continuity, naturalistic coloring, and complete fully-rounded objects with no unsightly artifacts or boundaries. In pushing the visualization closer to nature through "good design" aesthetics, the producers make the trail back to the original simulation harder and harder to recover. The visualization's voice, and the story it tells, grows more confident as it becomes distant and independent from the generating simulation, lending greater authority to its representation.

13.3 Spatial and Temporal (Dis)continuities

Within nature, most objects and systems that we observe and describe have a sense of spatial and temporal completeness, even systems such as weather that are described as being "chaotic." Trees and clouds, for example, are fully three-dimensional objects that one can walk (or fly) around without experiencing any unexpected disruptions or discontinuities in form, no visible tears or seams. The behavior of these objects and systems are also fairly predictable in time, at least, in the case of weather, for the short term.

The computer, on the other hand, requires that all simulated systems be defined as a collection of discrete, individual points, where the possibility of abrupt transitions due to boundary conditions or sharply changing data values always exists. It is an inherent property of computational simulations, as well as observed data, that they are spatially and temporally incomplete. Data is generated or collected on a one, two, three, or higher-dimensional grid, where for each point of the grid, a particular value of the data exists at a given moment in time. Each dimension, including time, is said to have a resolution, a finite value that describes the number of data points that exist in that dimension. Between these points of the grid, and outside the grid itself, the data values are not known.

The creators of visualizations are in a continuous battle against this "geometry of the grid [which] functions as a cartographic intervention, a fractured narrative mobilized by its own resistance to completion" [3]. For example, visualization producers use techniques such as interpolation to recover the data values in the hollow areas between grid points, an attempt to complete the data and minimize the grid's "intervention." Interpolation is used to create a sense of higher data resolution, a finer mesh in space and/or time that softens abrupt spatial transitions and reinforces the sensation that the simulation is flowing as a continuous process, rather than, for example, jumping in discrete chunks.

However, this sort of determination of the "in between" or "outside" data is always an approximation that assumes certain types of behavior in the hollow regions between grid and temporal points [4]. Data is not actually recovered in these empty regions, but merely blended across from existing data point to data point. More importantly, interpolation does not tell us anything new about the data, but rather operates strictly on an aesthetic level, making the visu-

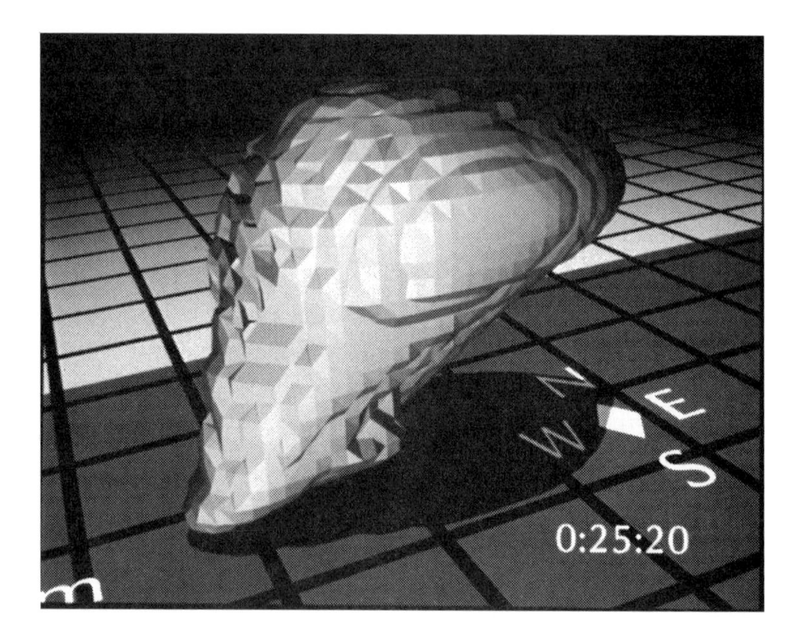

FIGURE 13.1. Storm cloud without smooth shading (NCSA 1993).

alization smoother and presumably more pleasurable and familiar. The visualization is constructed so as to compare favorably with the natural event that it is attempting to mimic. The closer to nature the visualization appears, the more convincing its own constructed narrative becomes.

Another common technique used in scientific visualization is "marching cubes," an algorithm that extracts surfaces of constant value, also known as isocontours or isosurfaces. Analogous to the elevation contour lines drawn on maps, these three-dimensional isosurfaces enclose data values that are higher than some specified threshold. The shapes that this algorithm creates are usually closed and visually coherent with a minimal amount of fragmentation.

However, if the underlying computational grid is too coarse, the isosurface generated can appear faceted (i.e. the surface appears to be made up of flat planes) (Figure 13.1). To correct for this, techniques such as smooth shading are used to soften the form's surface, blending the hard transitions in brightness (Figure 13.2). Faceting is considered an undesirable artifact within high-end computer graphics, and is generally interpreted as an indication of primitive software

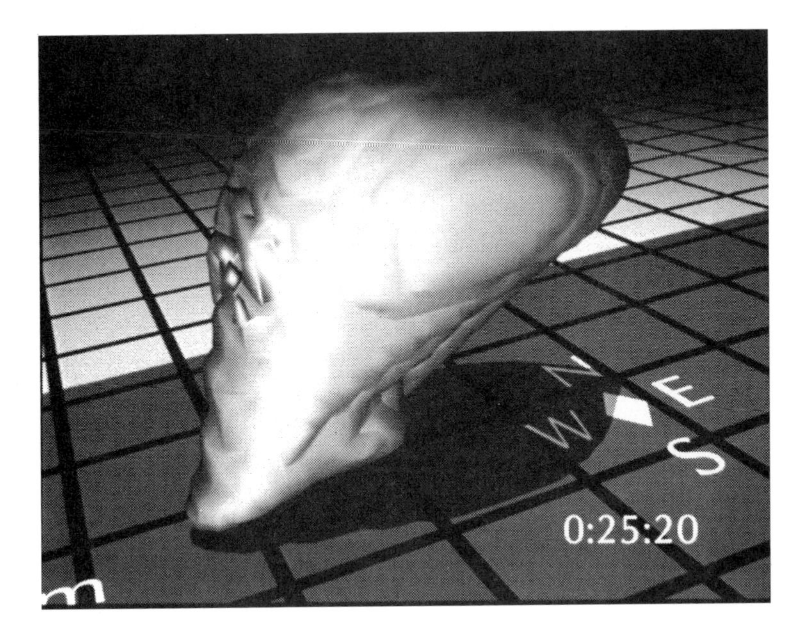

FIGURE 13.2. Storm cloud with smooth shading (NCSA 1989).

or hardware, or more damningly, a sign of aesthetic inexperience on the part of the user. Faceting is disruptive to a visualization's authority because it constantly reminds a viewer that the object on the screen is artificial and imperfect; the artifact disturbs the visualization's link to nature.

I do not mean to imply that simply including, for example, visual acknowledgments of the underlying grid of data is necessarily a more honest approach to scientific visualization. After all, here again an authoritative voice can be created through the evocation of apparent scientific honesty, generating a kind of "just the facts" tone. Rather, I am arguing that it is crucial for scientific visualizations to have embedded in them from the start an awareness and self-criticism of their own manufacturing. Hopefully, this will result in a visualization whose creation can be interrogated or at least traced by an audience, especially in situations where the graphics are being used as scientific or legal evidence. I will explore these issues further in the next two sections.

13.4 The Multiple Roles of the Visualization Expert

> The optical devices in question...are points of intersection where philosophical, scientific and aesthetic discourses overlap with mechanical techniques, institutional requirements and socioeconomic forces. Each of them is understandable not simply as the material object in question, or as part of a history of technology, but for the way in which it is embedded in a much larger assemblage of events and powers [5].

In order to understand some of the aesthetic decision-making that producers of visualizations perform, it is useful to look at their position within a network of social relationships. Because of visualization's tendency to be interdisciplinary, it is impossible to think of the practice as a clearly defined or pure activity. The creators of high-end visualizations exist within a network of tensions that lead them to continuously reconsider their performance, often within an environment that provides contradictory criteria. As intermediaries between disciplines that have traditionally viewed each other with some suspicion (such as academic scientists, artists and designers, corporate scientists, funding agencies, and the media), the visualization expert operates within a variety of different roles and attempts to simultaneously satisfy very different audiences [6].

13.4.1 Collaborator with Scientific Researchers

The first and usually primary role that a visualization expert fulfills is as a collaborator with the researchers that have computed or collected the raw data. Traditionally, these researchers have used computer graphics to assist in the basic analysis of their data, to produce animated versions of such familiar visual aids as line plots and bar graphs. Within a computer graphic environment, multiple parameters of the data can be shown simultaneously as they change over time. While enjoying this faculty of visualization, the researchers have often been suspicious of a presentation of their data which is not fairly unadulterated and functional.

However, more recently, as scientific researchers have become more knowledgeable and savvy about the power of high-end graphics, some have moved beyond the well-worn "raw vs. glitzy" debate that has plagued visualization from its start. These researchers recognize that higher quality renderings can provide a clarity of detail as well as a complexity in the sheer amount of information that can be displayed on the video or computer screen at one time.

Researchers also recognize the advertising potential of high-end graphics and use them to supplement grant proposals, or provide them to popular scientific television programs to expand their audience base beyond their research peers. As Dorothy Nelkin has written:

> Increasingly dependent on corporate support of research or direct congressional appropriations, many scientists now believe that scholarly communication is no longer sufficient to maintain their enterprise. They see gaining national visibility through the mass media as crucial to securing the financial support required to run major research facilities and to assuring favorable public policies toward science and technology [7].

Scientific visualization provides a seductive, self-contained vehicle for research scientists to distribute their work to a broader audience and potentially "gain national visibility" through exposure on broadcast and print media. This sort of distribution of a scientist's work requires the visualization to have a confident presentation of the data, internally consistent and convincing. Because of this need, the aesthetic decision-making process ends up playing an important role in the reception of the work. The more authoritatively the visualization can present itself, the better it fits into the media's general representation of science as a trustworthy and objective enterprise.

13.4.2 Artist

Another role that the creators of visualizations are sometimes positioned within is that of the filmmaker or artist. Here the visualization expert desires to transcend a simple examination of the data that is useful to the researchers and create a rich, preferably sublime (but not necessarily educational), experience for an audience that now extends far beyond a scientific one. In this vein, the visualizations have usually tended towards either an extremely realistic, trompe l'oeil style (landscapes and interiors being particularly popular), or else a highly abstract, patterned style, as for example with fluid flows and fractals. Very little art work that has its roots in visualization attempts to recontextualize the work in any way other than purely aesthetic (although there are certainly some exceptions to this) [8]. By not critiquing their own mode of creation, these art works implicitly celebrate the visualization process, in turn becoming attractive advertising material for the discipline.

13.4.3 Educator

The role that probably most strongly encourages visualization experts towards an aesthetic of nature is as creators of educational visualizations, where the work is specifically designed for use on popular science programs and exhibits at science museums. Here the desire is to present animations and still images with clear, self-explanatory narratives that require a minimum of voice-over or printed text. The goal is to produce images that are instantly recognizable to as many people as possible, and thereby become persuasive exactly because they are so recognizable and "close to nature." By minimizing the difference between the simulation and some shared natural event, the need to convince an audience of the validity of a particular simulation or representation disappears. After all, the logic seems to go, if the underlying simulation was not accurate, then the graphics derived from the simulation would not so closely resemble ones conception of the "actual event." This issue can become somewhat more problematic when one recognizes that the resemblances may have been manufactured through layers of aesthetic decision-making, and are not necessarily an integral part of the underlying simulation.

For educational purposes, it makes a certain amount of pedagogical sense to employ realism in image-making, so that, to quote John Tagg, "the process of production of a signified through the action of a signifying chain is not seen...the product is stressed, the production is repressed" [9]. This allows the message to travel to the audience seemingly unencumbered with extraneous, possibly distracting, issues. But one of the problems that arises with this move towards "straightforward" communication is that the audience's ability to interrogate the visualization vanishes in the details of mimicking nature's subtleties. The naturalistic aesthetic becomes desirable to the visualization expert, scientific researchers, and their audiences exactly because it seems to operate so cleanly and without resistance.

However, in the end there is the danger that education can fall victim to pleasure.

13.4.4 Manufacturer of Evidence

Another recent application of visualization that grants it an entirely new level of participation within our society is as scientific evidence in the courtroom (e.g., accident re-creations) and policy making arenas such as congressional committees (e.g., visualization of air pollutants over a city). Here the visualization must walk a very fine line between appearing convincing enough so that it will be admissible as evidence, and appearing over-produced, thus risking a confusion with advertising [10].

The attitude of lawyers who have used animations in the courtroom has been that the computer graphics are simply clarifying expert witness testimony, that everything shown in an animation has a one-to-one correspondence with what an expert says, which, ideally, has a one-to-one correspondence with the actual event(s) [11]. The problem lies exactly in this desire to simplify complicated issues and circumstances, which not coincidentally parallels what many researchers are striving for in scientific visualization. The ability for one side of an argument to present animations to the jury as evidence creates the sense that this particular side has an insight to the actual events that the other side does not. Even though the animation is simply a construction, a visually presented hypothesis, the sense of insight is concretized through the power of the televised graphics. For example, following the conviction in the murder trial *People v. Mitchell*, the victorious District Attorney John Posey commented that without the computer animated reconstructions, the outcome of the trial would have been different [12].

The producers of these animations, of course, are fully aware of the power of the graphics on the jury, as well as their contrived construction:

> [J]urors tend to believe what they see. This technology keeps the jurys attention by simplifying the material and by giving them little bursts of information...were looking to develop a complete graphics strategy that will focus the jury precisely on the facts and arguments on which we want them to base their verdict [13].

In considering the process of creating these courtroom visualizations, some interesting questions come to mind. For example, if in the re-creation of an accident with computer graphics it appears that the product the graphics are trying to defend was in fact defective in some way, would the visualization process be discontinued? Or would a different tact be taken to change the conclusion one might draw from a computer graphic animation? It is very easy to change *this* slight detail of a motion of an object, or make this hill *just* an inch shorter, or even just move the camera *slightly* over here, to completely change ones impression of the event. Although lawyers and judges seem confident that the process can be monitored and the accuracy of a computer graphic re-creation determined [14], the levels of detail and complexity involved in its creation make this a difficult task indeed. A visualization process cannot be cross-examined in the same way that an expert witness can.

Although the cost associated with the production of courtroom graphics continues to drop, it is still important to point out that the expense of creating these animations is by no means insignificant. This high cost of production favors the party that has the fiscal and logistical resources to commission the re-creation, which undoubtedly will usually be the corporation involved in the suit as opposed to the individual. Even when the price for courtroom animations becomes truly "democratic," the corporation will still have the edge by being able to commission the latest, most visually sophisticated technologies.

I have described these social roles of the visualization expert because it is important to recognize that scientific visualization is not simply an unmediated, direct translation from simulation or observation to image, but rather is the product of a group of people that make a complicated set of decisions based on such things as aesthetic taste, perceived expectations of their audience, history of the medium (e.g. commercial animation), institutional and corporate pressures, funding sources, and desires for fame. In writing about the creation of maps, John Pickles describes very much the same sort of process:

> The map is a purposive cultural object with reasons behind its construction and values associated with its reading. To suggest otherwise is to fail to see its status as made object... In mediating the transformative processes of abstraction, reduction, thematization and idealization,

the cartographer selects, sifts and emphasizes this or that aspect of the world under construction [15].

13.5 Case Studies

As a way of illustrating some of the ideas presented in this paper, I would like to analyze three scientific visualizations produced at NCSA. Having personally worked on the second and third visualizations described here, I feel that I can bring insight to the decision-making that went into their creation.

13.5.1 Simulated Tornadic Storm Development

In 1987, atmospheric research scientists at NCSA collaborated with members of the NCSA Scientific Visualization Group to produce a high-end visualization that depicted the development of a simulated tornadic thunderstorm. The goal of this visualization, as with most, was to provide a visually coherent glimpse into a complex, dynamically changing three-dimensional grid of data. To render the scene, the team used Wavefronts Advanced Visualizer, a high-quality 3-D graphics software package originally designed for use in commercial animations.

The animation that resulted from this collaboration was an impressive sequence focusing on a dynamic white cloud racing across a plowed field in the late afternoon light (Figure 13.3). Not surprisingly, the visualization attracted much attention and acclaim from the scientific research and computer graphics communities, as well as from the media. This animation became one of the early icons of scientific visualization, a testament to its potential power and breadth of application.

However, the visualization also attracted some criticism from researchers because of the depiction of the storm as developing over an irregularly surfaced, plowed dirt field. The computer simulation of the cloud that the researchers used to generate the data was not calculated over such a grooved surface. In computational simulations, data located at a boundary that is rough will behave differently than data at a boundary that is perfectly smooth. This difference in behavior at the boundary can potentially change the outcome of the entire data set, steering the simulation toward a different end.

FIGURE 13.3. Simulated Tornadic Thunderstorm Development (NCSA 1987).

By placing the simulated storm over a grooved surface, the critics argued, the visualization was sloppy in its contextualization of the underlying simulation.

In addition to this issue, what I find intriguing is the mistaken identification of the cloud that arises from a misrepresentation of scale and from a similarity in form. Looking at the size of the grooves in the field relative to the size of the cloud, as well as the shape and dynamics of the cloud, one is compelled to read the white conical shape as a simulated tornado [16]. But this simulated object is not a tornado at all, but rather the large water particles of a storm cloud, a cloud much larger than a tornado itself. A tornado would be but one very small detail in a cloud of this size: represented to scale, a tornado would about fit in one of the grooves in the "field." Due to the spatially coarse computational grid, such a detail could never even be observed in this particular simulation [17].

Because of the ability of the cloud to evoke something outside itself, to evoke something it is in fact not, our ability to understand the intent of the original data is severely compromised. The computer graphic object failed because it was not visually faithful to

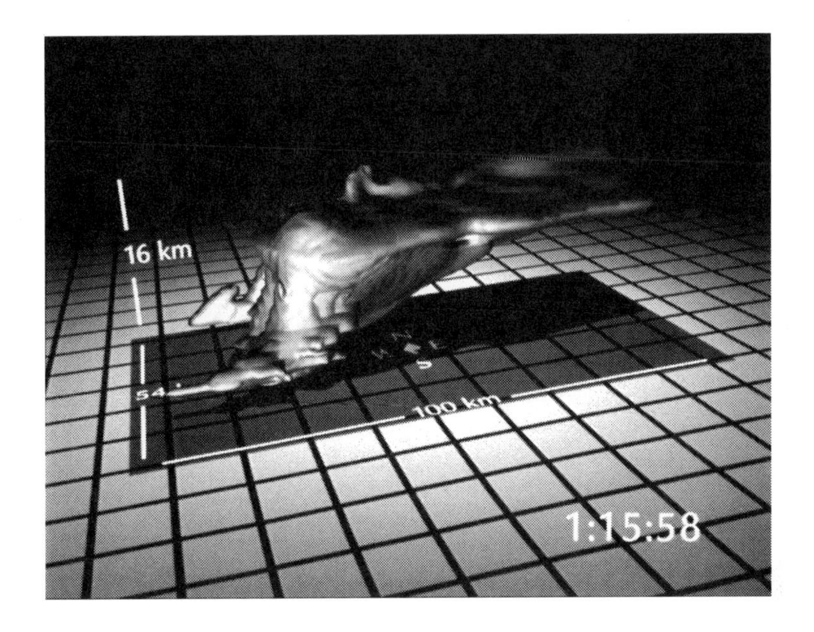

FIGURE 13.4. Study of a numerically modeled severe storm (NCSA 1989).

what it was supposed to represent, a large cloud, so it was made into something else that it more closely resembled, a tornado. By transforming the context that the simulated object was presented within, the visualization process reached back and in effect redefined the simulation. As Simon Penny has observed,

> [A]s the verisimilitude of the representation increases, the simulation can be approached in a self-contained game-playing mode, where the goal is to attain a maximum score in the terms of the model, complete separation being achieved from the corporeal situation the model purports to represent....The simulation is taken to be true on the basis of its coherent structure of representation, while the database that provides the source statistics for the representation may be inaccurate or incomplete [18].

Or, we may add, misplaced.

FIGURE 13.5. Live footage from study of numerically modeled severe storm (NCSA 1989).

13.5.2 Study of a Numerically Modeled Severe Storm

In 1989, two years after the first video, new simulations of tornadic thunderstorms were used to create another visualization, with the hope that this one would be more careful in its exploration and contextualization of the data. The simulation is cycled through a total of eight times in the video, each iteration studying a different aspect of the data (which includes heavy and light water particles, wind velocity, temperature, and vorticity) [19]. The visualization team made a conscious decision to place the developing storm over a stark black and white grid in order to mimic what we imagined to be a sterile "computational laboratory," a virtual space where one performed scientific experiments, but instead of using beakers, test tubes, and Bunsen burners, one used computers (Figure 13.4). This visually austere floor was a deliberate move away from the problematically rich, plowed field of the earlier video. In addition, the new video opens with dramatic live footage of a tornado crossing a road right in front of a car full of "storm chasers" (Figure 13.5). The grainy, smeared

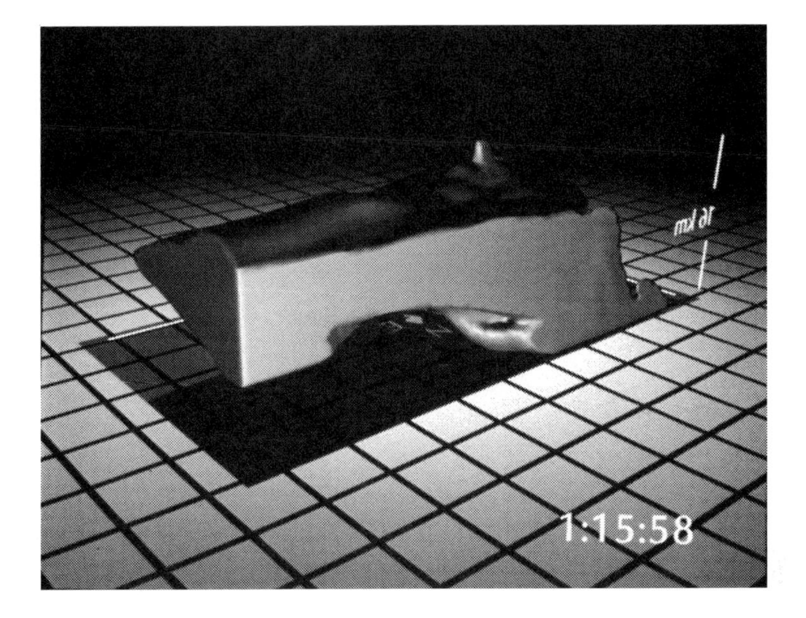

FIGURE 13.6. Back view of same cloud shown above (NCSA 1989).

quality of this footage contrasts sharply with the computer-clean visualizations that follow, positioning the latter even further within the simulation space of the computational lab.

Given this relative upfrontness about the computational origins of the storm, it is interesting to note that there were other aesthetic concerns we were not willing to sacrifice. In particular, in seven minutes of computer generated imagery, the camera never travels outside of a 90 degree region south and west of the developing storm, so that in the end only half the cloud is ever seen. The reason for this limited travel by the camera is simple enough: the simulated cloud runs into the boundaries of the computational domain on the northern and eastern edges. In the latter part of the simulation, these two boundaries create rather unappealing flat faces, almost as if someone had sliced portions of the cloud away (Figure 13.6) [20].

One could argue that there is good reason for the camera to never stray "behind" the storm. The surface there is not very informative in terms of analysis of the data. Given the richness and complexity of the fully rounded southern face, why spend time looking at a flat wall, an unfortunate but very typical artifact in computational

FIGURE 13.7. Visibility analysis though animation of a backhoe work cycle (NCSA 1991).

data? However, one could also argue that by never once explicitly showing the computational boundaries of the cloud, we were moving the storm away from its computational origin and aligning it more closely with an ideal referent located in nature. By never showing the violent disruption to this computer generated form, a desire to minimize the artificialness of the simulation and preserve the integrity of the cloud seems to be at work. The visualization team used the computer graphic camera to limit the sight of the audience and as a result, position the visualization of the storm at a more comfortable distance from its computational origins [21].

13.5.3 Visibility Analysis Through the Animation of a Backhoe Work Cycle

In 1991, I worked on a project with one of NCSAs industrial partners, Caterpillar Inc., that I believe made an attempt at including the construction of the visualization as an integral part of the visualization itself. The goal of the project was to explore different designs of backhoes and assess the visibility of the vehicle operator from

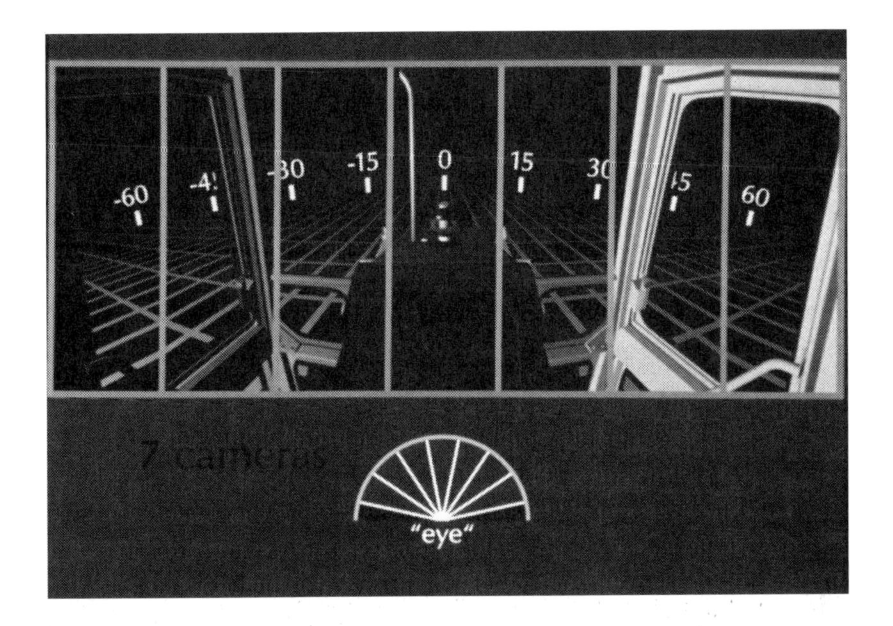

FIGURE 13.8. Composite view created with seven cameras (NCSA 1991).

within the cab, to discover potential problems such as blind spots and poor lines of sight to the buckets and surrounding environment (Figure 13.7) [22].

One of the central concerns that we addressed was how to simulate the viewpoint of the operator from within the cab. A default camera in a computer graphic environment has a horizontal angle of view of approximately 45 to 50 degrees. The view from within the cab with this sort of camera definition is not sufficient as a representation of an operator sitting within a real cab, since with peripheral vision humans can see close to 180 degrees horizontally.

By redefining the camera to create a wider angle of view, the resultant image gave a better sense of being positioned within the cab. The interior of the cab was now visible, as was approximately a sweep of 155 degrees of the surrounding environment. However, the handling of space across the image plane was very inconsistent, with dramatic stretching and compressing, much like a fisheye lens. This uneven handling of space made this camera definition unacceptable as well.

The solution we arrived at for modeling the operators viewpoint utilizes a constructed composite view made from seven separate cameras, which provides the width of the wide-angle lens while at the same time minimizing spatial distortions (Figure 13.8) [23]. The 155 degree horizontal view of the wide-angle lens is now broken up into seven smaller views of approximately 22 degrees each. These seven views are individually rendered and then assembled on the image plane to create a composite image covering the same angle of view as the wide-angle lens.

In the video, the construction of this faceted "eye" is shown explicitly. In general, the camera in a visualization is constantly influencing what one is seeing, yet it is generally an unacknowledged participant; its presence is naturalized and forgotten. In this particular visualization, this apparent passivity of the camera could no longer be either sustained or tolerated since the camera was in fact creating the data. It seemed to us imperative to foreground the artificialness of the multi-camera viewpoint.

This notion of cameras creating the data was made even more explicit with the addition of another camera window that literally measures the percentage of the bucket that is visible and hidden at a given moment in the work cycle (Figure 13.9, lower left corner). The colors of the bucket and the cab in this window are no longer tied to the objects themselves. Rather, the objects are broken down into patches of color where the hue is dependent on whether a particular region of the object is visible to the operator, is obscured, or is doing the obscuring. Color functions here as a carrier for data as opposed to the re-creation of an object's actual appearance.

By combining the seven camera views, the bucket blockage window, and a graph and numerical display of the time dependent data from the blockage window all on one NTSC image, the viewer has a number of information windows into the simulation space simultaneously (Figure 13.9). The co-existence of these different modes of "seeing" underscores the incompleteness of any one representation of vision. This fracturing of a monolithic viewpoint into a number of different possible should help the audience to recognize the manufacturing process of the data.

FIGURE 13.9. Multiple representations of an operator's visibility from within the cab (NCSA 1991).

13.6 Conclusion

Scientific visualization is playing an ever increasing role in the mediation of complex information between researchers and the general public. The animations are often striking and seductive, which, in of itself, is not necessarily a bad thing. However, it is important for scientific researchers and visualization producers to recognize that seemingly benign decisions made in the creation of a video or interactive system can manufacture misleading narratives and associations that are difficult for an audience to interrogate. It is the responsibility of visualization producers to create works that do not attempt to mask their artificialness, but rather present this property as an integral component of the story-telling.

13.7 References

[1] Friedhoff, R. and W. Benzon (1989). *Visualization: The Second Computer Revolution.* New York: Harry Abrams, p 12.

[2] See Deitchs essay in J. Deitch and D. Friedman (1990). *Artificial Nature.* Athens: Deste Foundation for Contemporary Art.

[3] Leone, H. and J. Macdonald (1992). Passio perpetuae. In Crary, J. and Kwinter, S.(Eds.), *Zone 6: Incorporations.* New York: Zone, p. 597.

[4] For example, that the data changes linearly across space and/or time.

[5] Crary, J. (1990). *Techniques of the Observer: On Vision and Modernity in the Nineteenth Century.* Cambridge, MA: MIT Press, p. 8. Although Crary is referring here to "optical devices" of the nineteenth century (stereoscopes, phenakistiscopes, zootropes, kaleidoscopes), his point can certainly be extended to the computer graphic/scientific visualization "optical device." I am also extending the "device" to include the human producer of the visualization, who operates as a cybernetic-organism (cyborg) in relationship with the hardware and software of the computer.

[6] In this section of the paper I will refer to the producer of visualizations as a "visualization expert." I use this term partly because it distinguishes the producer from the scientific researcher, partly because no other term is really satisfactory. Other possible labels would be "animator," "scientific animator," and "visualizer."

[7] Nelkin, D. (1987). *Selling Science: How the Press Covers Science and Technology.* New York: W. H. Freeman and Co.,p. 133.

[8] See, for example, "Mandel Zoom" by Booker Bense, the phscolograms of (Art)^n, and "In Search of the Fingerprints of God" by Mark Bajuk and Matthew Arrott. Although not based on computer graphic imagery, "Fauna" by Joan Fontcuberta and Pere Formiguera explores the fabrication of an authoritative scientific narrative, by using seemingly authentic field notes and photographs to document nonexistent, fantastical species.

[9] Tagg, J. (1988). *The Burden of Representation: Essays on Photographies and Histories.* Amherst, MA: The University of Massachusetts Press, p. 99.

[10] Stix, G.(1991). Seeing is believing: A picture may be worth a million-dollar settlement. *Scientific American,* 265, (December), 142.

[11] See, for example, Dilworth, D. (1992). Computer animations reach criminal court. *Trial,* 28, (September), 26; see also Bulkeley, W. (1992). More lawyers use animations to sway juries. *Wall Street Journal,* (18 August), B1.

[12] Posey, J. quoted in Chu, A. (1992). Sex, lies and computer animation. *Security Management,* 36, (October), 13.

[13] Gregory Mazares, president of the graphics unit at Litigation Sciences Inc., quoted in "Computer Graphics Aiding Jurors Recall," *New York Times,* (24 November 1989), B-27.

[14] See Bulkeley, B1; Dilworth, 26; and Kreiger, R. (1989). New dimensions in litigation: Computer-generated video graphics enter the courtroom scene. *Trial,* 25, (October), 74-76.

[15] Pickles, J.(1992). Texts, hermeneutics and propaganda maps. In Barnes, T. and Duncan, J. (Eds.), *Writing Worlds: Discourse, Text and Metaphor in the Representation of Landscapes.* New York: Routledge, 221. Emphasis in original.

[16] A number of articles and books that have reproduced the image have described the object as a tornado. See, for example, Friedhoff, 162-163.

[17] The coarseness of the computational grid is due to the limitations in memory, disc space, and compute power of supercomputers at that time.

[18] Penny, S. (1993). Disentangling utopian dreams. In Fifield, G. and Wallace,B. (Eds.), *The Computer is Not Sorry,* exhibition catalog. Boston: The Space, 2.

[19] For details on the simulation, see Wilhelmson, R. et al. (1990). A study of the evolution of a numerically modeled severe storm. *The International Journal of Supercomputing Applications,* 4(2), 20-36.

[20] The image reproduced here is not shown in the video; it is an image I created specifically for this paper.

[21] Although this problem may seem to be unique to visualization within a video format, I have come across at least one example of interactive scientific virtual reality where the user's mobility was subtly limited to discourage travel into regions apparently deemed undesirable by the creators.

[22] See Bajuk, M. (1992). Camera evidence: Visibility analysis through a multi-camera viewpoint. In Alexander, J.R. (ed.), *Visual Data Interpretation Proc. SPIE 16,* 68, 61-72.

[23] The frames around the seven windows are temporarily displayed in the video to clarify the construction of the composite camera. They are removed during the simulated work cycle.

14

Crossroads in Virtual Reality

Michael Heim[1,2]

14.1 Introduction

The public knows Virtual Reality (VR) as a synthetic technology combining 3-D video, audio, and other sensory components to achieve a sense of immersion in an interactive, computer-generated environment. Behind the quiet facade of this definition, however, burns a hot debate about what VR is or can be. The debate is sparked by the clash of two opposing directions in VR development, each arising from a distinct philosophy of representation and communication. Each view conceives differently what it is that interactive computer graphics represents and how human beings might benefit from it. One line of thought focuses on communication through unencumbered telepresence (e.g. Myron Krueger, the Mandala system, the CAVE from EVL, and MIT's ALIVE), while the other concentrates on the design of occlusionary 3-D graphic interfaces with multi-sensory simulation based on a stereoscopic head-mounted display or HMD (UNC, Wright-Patterson, NASA, and the HIT Lab). By tracing the underlying threads of these two developments, we can locate the reality theories behind them. Examining these theories, we get a better look at the consequences and trade-offs of each approach. The outside observer usually sees the two developments as crossroads or opposing trends. The philosophical insider, however, realizes that the two trends might one day blend into a complementary system that wards off some of the hazards of a one-sided VR technology. Since VR achieves the greatest possible interiorization of symbolic life in our evolutionary history, we should envision the complementary nature of the two types of VR interface. I suggest

[2]Education Foundation of the Data Processing Management Association, 2041 San Anseline, No.4, Long Beach, CA 90815

a model of complementary VR combining both perceptive and apperceptive graphics. Projection graphics becomes a decompression chamber in which a T'ai Chi master helps the VR user re-integrate mind and body as the user returns from HMD cyberspace.

14.2 The Meanings of Virtual Reality, Virtual Worlds, Virtual Environments

Before examining the two main developments in VR graphics, I want first to mention the general context in which these two developments currently exist. Virtual Reality still functions as an umbrella concept for several related research and commercial products, all of which claim public attention. While two main contenders will, I think, eventually clear the deck and eliminate the other more loosely associated brands of VR, I want, nevertheless, to describe several meanings of virtual reality now used by observers and even by some computer scientists. I sketched the spectrum of the meanings of virtual reality some time ago in Chapter 8 of *The Metaphysics of Virtual Reality* [1], but I will re-establish the umbrella concept here before going on to analyze the two main contenders.

Webster's dictionary tells us that the "virtual" in virtual reality means that something virtual is real "effectively but not formally." So when a computer graphic makes an entity present to us so effectively that we might just as well have it before us, then the graphic becomes virtually that entity. The graphic then provides a virtual reality. The dictionary definition of virtual goes back to a verbal distinction forged with great precision by Duns Scotus, a scholastic philosopher of the late Middle Ages. But when we look at what various researchers mean today by virtual reality, we are promptly ejected from medieval scholastics and are thrown back on the novelty of a concept we must redefine here and now.

What does it mean specifically for a computer graphic to make something present "in effect but not in fact"? Of course, the virtual presence of objects and persons requires more than a visual representation. Other sensory modes are needed. The eyes need help from the other senses to achieve that special quality of virtual presence. Achieving multi-sensory experience synthesized by the computer is a complex challenge with many possible solutions. And here is where the meaning of VR branches out into several meanings, each going off in a slightly different direction. Each direction brings in a different aspect of virtual reality. At least seven different concepts currently fall under the general umbrella of VR. The seven include: Simulation, Interaction, Artificiality, Immersion, Telepresence, Full-Body Immersion, and Networked Communications[1]. These different concepts set the field for the two main emerging camps that have different philosophical roots and that fervently disagree about how to construct virtual worlds.

14.2.1 Simulation

Computer graphics have such a high degree of realism today that the sharp images evoke the term virtual reality. As sound systems were once praised for their high fidelity, present-day imaging systems now deliver "virtual reality." The images have a shaded texture and light radiosity that pulls the eye into the flat plane with the power of a detailed etching. Landscapes produced on the GE Aerospace "visionics" equipment, for instance, are photorealistic real-time texture mapped worlds through which users can navigate. These graphic dataworlds spring from the context of mission rehearsal and training in military flight simulators. Now researchers are applying these graphic techniques to medicine, entertainment, and education.

The realism of simulations applies to sound as well. Three dimensional sound systems control every point of digital acoustic space. 3-D audio exceeds earlier sound systems to such a degree that audio becomes a necessary part of virtual reality.

14.2.2 Interaction

Some people consider virtual reality to be any electronic representation with which we can interact. Cleaning up our computer desktop, we see a graphic of a trash can on the computer screen, and we use a mouse to drag a junk file down to the trash can to dump it. The desk is not a real desk but we treat it as if it were, virtually, a desk. The trash can is an icon that triggers a deletion program, but we use it – in the existential semantics of the term – as a virtual trash can. And the files of bits and bytes we dump are not real (paper) files but function virtually as files. These are virtual realities. What makes the trash can and the desk different from cartoons or photos on TV is that we can interact with them as we do with metal trash cans and wooden desktops. The virtual trash can does not have to fool the eye in order to be virtual. Illusion is not the issue. The issue is how we interact with the trash can as we go about our work. The trash can is real in the context of our absorption in the work, yet outside the computer work space we would not speak of the trash can except as a virtual trash can. The reality of the trash can comes from its handy place in the world woven by our engagement with a project. It exists through our interaction.

Defined broadly, virtual reality sometimes stretches over many aspects of electronic life. Beyond computer-generated desktops, it includes the virtual persons we know through telephone or computer networks. It includes the entertainer or politician who appears on television to interact on the phone with callers. It includes virtual universities where students attend classes on-line, visit virtual classrooms, and socialize in virtual cafeterias.

14.2.3 Artificiality

As long as we are casting our net so wide, why not make it cover everything artificial? Many people, on first hearing "virtual reality," respond immediately: "Oh, sure, I live there all the time." By which they mean that their world is largely a human construct. Our environment is thoroughly geared, paved, and wired – not quite solid and substantial. Planet earth has become an artifice, a hybrid product of natural and human forces combined. Nature itself, the sky with its ozone layer, no longer escapes human influence. And our public life has everywhere been computerized. Computer analysis of purchasing habits tells supermarkets how high and where to shelve the Cheerios. Advertisers boast of "genuine simulated walnut."

Years ago, Daniel Boorstin remarked that the increasing congestion of city traffic, the parking problem, and the lengthy holding patterns over airports have made our television screens a superior way of getting somewhere. When it comes to public events, Boorstin wrote[2], "you are often more there when you are here than when you are there." Much of what we experience of public life comes through TV, so much of our public life already occurs in virtual reality.

But once we extend the term "virtual reality" to cover everything artificial, we lose the force of the term. When a word means everything, it means nothing. Even the term "real" needs an opposite.

14.2.4 Immersion

Many people in the VR industry prefer to focus on a specific hardware and software configuration. This is the model Ivan Sutherland set for virtual reality, which continues through the work of Scott Fisher, Tom Furness, and Fred Brooks. Before the head-tracking stereoscopic display existed, there was no term "virtual reality," since no hardware or software existed to claim the name. Only after Jaron Lanier dubbed the hardware-software configuration "virtual reality" did the concept emerge as a cultural phenomenon with an industry building around it.

The specific hardware first called VR combines two small 3-D stereoscopic optical displays or "eyephones," a Polhemus head-tracking device to monitor head movement, and a Dataglove or hand-held device to allow feedback so the user can manipulate objects perceived in the artificial environment. Audio with 3-D acoustics can enhance the illusion of being submerged in a virtual world. The illusion is one of immersion, of feeling "you are there" amidst the computer-generated entities. The term "telepresence" often means this special kind of being there, sometimes including an additional graphic that represents the user's self as a cyberbody or telepresent first-person actor.

Virtual reality, on this view, means a system for sensory immersion in a virtual environment. Such systems, known primarily by their head-mounted displays (HMD) and datagloves, were first popularized by Jaron Lanier's company, VPL (Virtual Programming Language) Incorporated, a company that shipwrecked on a bad investment portfolio. The HMD popularized by VPL cuts off visual and audio sensations from the surrounding world and replaces them with computer-generated sensations. In a similar hardware system at the University of North Carolina invented by Fred Brooks, the body moves through artificial space using feedback gloves, foot treadmills, bicycle grips or joysticks.

A prime example of immersion comes from the U.S. Airforce, where some of this hardware was first developed for flight simulators. The computer generates much of the same sensory input a jet pilot experiences in an actual cockpit. The pilot responds to the sensations by, for instance, turning a control knob, which in turn feeds into the computer which again adjusts the sensations. In this way the pilot can practice or train without leaving the ground. To date, commercial pilots can upgrade their licenses on certain levels by putting in a specified number of hours on a flight simulator.

Computer feedback may do more than re-adjust the user's sensations to give a pseudo-experience of flying. The feedback may also connect to an actual aircraft, so that when the pilot turns a knob, a real aircraft motor turns over or a real weapon fires. The pilot in this case feels immersed and fully present in a virtual world, which in turn connects to the real world. This linkup between graphics scenario and actual landscape constitutes either telepresence robotics or augmented reality.

Augmented reality is when the heads-up display in the cockpit permits the pilot to view the real landscape behind the virtual images. When you are flying low in an F-16 Falcon at supersonic speeds over a mountainous terrain, the less you see of the real world, the more control you have over your aircraft. A virtual cockpit filters the real scene and re-presents a more readable world. In this sense, VR preserves the human significance of an overwhelming rush of split-second data. In such cases, the simulation is an augmented rather than a virtual reality.

The arcade game offshoots of this technology, such as the Virtuality series from the UK's W Industries are only simplified versions of the applications going on in molecular biology (docking molecules by sight and touch), airflow simulation, medical training, architecture, and industrial design. Boeing Aircraft plans to project a flight controller into virtual space, so that the controller floats thousands of feet above the airport, looking with an unobstructed view in any direction (while actually seated in a datasuit on the earth and fed real-time visual data from satellite and multiple camera viewpoints).

A leading model of this research has been the workstation developed at NASA-Ames, the Virtual Interface Environment Workstation (VIEW). NASA uses the VIEW system for telerobotic tasks, so that an operator on Earth feels immersed in a remote but virtual environment and can then see and manipulate objects on the Moon or Mars through feedback from a robot.

Immersion research concentrates on a specific hardware and software configuration. The immersive tools for pilots, flight controllers, and space explorers are a much more concrete meaning of VR than the vague generalization "everything artificial."

14.2.5 Telepresence

Robotic presence adds another aspect to virtual reality. To be present somewhere yet present there remotely is to be there virtually (!). Virtual reality shades into telepresence when you are present from a distant location – "present" in the sense that you are aware of what's going on, effective, able to accomplish tasks by observing, reaching, grabbing, moving objects as if close-up and with your own hands. Defining VR by telepresence nicely excludes the imaginary worlds of art, mathematics, and entertainment. Robotic telepresence brings real-time human effectiveness to a real world location without there being a human in the flesh at that location. Mike McGreevy and Lew Hitchner walk on Mars, but in the flesh they sit in a control room at NASA-Ames.

Telepresence medicine places doctors inside the patient's body without major incisions. Medical doctors, like Colonel Richard Satava and Dr. Joseph Rosen, routinely use telepresence surgery to remove gall bladders without traditional scalpel incisions. The patient heals from surgery in one-tenth the usual time because telepresence surgery leaves the body nearly intact. Only two tiny incisions allow the introduction of laparoscopic tools. Telepresence allows surgeons to perform specialist operations at distant sites where no specialist is present in the flesh.

Allowing the surgeon to be there without being there, telepresence is a double-edged sword, so to speak. Telepresence permits immersion where the operator gains great control over remote processes. But, at the same time, a psycho-technological gap opens up between doctor and patient. Surgeons complain of losing hands-on contact as the patient evaporates into a phantom of bits and bytes.

14.2.6 Full-Body Immersion

About the same time head-mounted displays appeared, a radically different approach to VR was emerging. In the late 1960s, Myron Krueger, often called "the father of virtual reality," began creating interactive environments where the user moves without encumbering gear. This type of VR sometimes goes by the name "projection VR."

Krueger's come-as-you-are VR uses cameras and monitors that read the human body. The cameras input gesture into the computer so the computer graphic image of the user's body can adjust in real time and its projected image can then interact with the computer-generated graphic environment. The user's hands can manipulate graphic objects on a screen, whether text or pictures, just as in the occlusionary HMD approach. But the interaction of computer and human takes place without encumbering the body or without limiting the primary body's autonomy. The burden of input rests with the computer, and the body's free movements become text for the computer to read. Cameras follow the user's body, and computers synthesize the user's movements with the artificial environment.

I see a floating ball projected on a screen. My computer-projected hand reaches out and grabs the ball. The computer constantly updates the interaction of my body and the synthetic world I see, hear, and touch.

The focus of this VR immersion resides in the projected images. Immersion refers to an act of imaginative physical involvement as much as forgetting the primary world. The user becomes telepresent in the images of the graphic environment as they intermingle with the images of the real-time changes in the user's bodily movements. Consequently, the user does not submit to the virtual environment with the same slavish tunnel-vision induced by the HMD. For this type of immersion, the name "unencumbered VR" seems appropriate, even though it remains a negative definition by the contrast with the "encumbered" or HMD experience. Theorists use the positive name "projected VR" to designate Krueger's line of work.

In Krueger's Videoplace, people in separate rooms relate interactively by mutual body-painting, free-fall gymnastics, and tickling. Krueger's Glowflow, a light-and-sound room, responds to people's movements by lighting phosphorescent tubes and issuing synthetic sounds. Another environment, Psychic Space, allows participants to explore an interactive Maze where each footstep corresponds to a musical tone, all produced with live video images which can be moved, scaled, rotated, without regard to the usual laws of cause and effect.

14.2.7 Networked Communications

Pioneers like Jaron Lanier accept the immersion model of virtual reality but add equal emphasis to another aspect they see as essential. Because computers make networks, VR seems a natural candidate for a new communications medium. The RB2 System ("Reality Built for Two") from VPL highlights the connectivity of virtual worlds.

A virtual world, in this view, is as much a shared construct as the telephone. Virtual worlds, then, can evoke unprecedented ways of sharing, what Lanier calls "post-symbolic communication." Because users can stipulate and shape objects and activities of a virtual world, they can share imaginary things and events without using words or real-world references.

In this view, communication can go beyond verbal or body language to take on magical, alchemical properties. A virtual world maker might conjure up unheard of mixtures of sight, sound, and motion. Consciously constructed outside the grammar and syntax of language, these semaphores defy the traditional logic of verbal and visual information. VR can convey meaning kinetically and even kinesthetically. Such communication will probably require elaborate protocols as well as lengthy time periods for digesting what has been communicated. Xenolinguists will have a laboratory for experiment when they seek to relate to those whose feelings and world-views differ vastly from their own.

Many researchers are now turning their attention to the development of on-line networked communities. Even under the constraints of alphabetic symbols, virtual communities spring into existence by using visualization and other imaginative techniques. Our culture seems to be in the process of honing its skills at creating virtual worlds based on game-like role-playing and participatory imagination.

Drawing on this brief overview of current meanings, we can easily blend all seven meanings so that they coexist under a single umbrella definition:

An artificial simulation can offer users an interactive experience of telepresence on a network that allows users to feel immersed in a communications environment.

The amalgam only slightly revises the meanings so they blend smoothly. No direct contradiction arises. Yet one of the seven meanings does not fit as nicely as the others.

Full-body immersion or projection VR stands out. Projection VR stands out distinct and even counter to the HMD approach. These two types of VR interpret immersion differently. In the most popularly and widely known VR, the HMD (goggles & gloves) achieves immersion by occluding the surrounding environment. Projected VR, on the contrary, preserves bodily freedom while generating a graphic cyberbody to merge with other computer-generated entities. In fact, projected VR seems to provoke bodily movement in the user.

Yet Krueger's projection VR has received little recognition by the industry. Especially during the early phase of popularization, in the late 1980s, the projection type of VR went largely unacknowledged, and Myron Krueger had to struggle to get a hearing for his systems. In the late 1980s, projection VR failed to capture anything like the attention – or the research funds – given to the HMD systems. Funding at NASA and military research labs fell largely into the hands of those working on systems with head-mounted displays and datagloves, despite the obvious limitations of the current stereoscopic LCD displays. (Krueger was first to point out that resolution of the LCD graphics in the 1990 HMD displays bring the user's visual acuity down to the level of "legally blind.")

In the early 1990s, however, interest rekindled in projection VR. The Mandala system, drawing on many of Krueger's ideas, achieved notable success in sports, dance, and entertainment applications. Other researchers boosted projection VR to new levels. In 1992, the CAVE at the Electronic Visualization Lab of the University of Illinois at Chicago introduced surround-screen projection-based techniques to create an entire room where the user could explore VR with unencumbered body. (CAVE is a recursive acronym for: "CAVE Audio Visual Experience, Automatic Virtual Environment." The name also derives from Plato's simile of the Cave in The Republic.) The CAVE's researcher, Carolina Cruz-Neira, brought Krueger's ideas to a new fruition under the tutelage of Krueger's former colleagues, Thomas DeFanti and Daniel Sandin. Cruz-Neira continues to develop further ways of integrating high-speed computation applications in the CAVE environment. She is also developing libraries of virtual environments that she can install in a matter of hours.

The 1993 ALIVE demonstration at SIGGRAPH '93 in Anaheim used projection VR. ALIVE is an "Artificial Life Interactive Video Environment" designed at the MIT Media Lab by Pattie Maes and Bruce Blumberg. The Media Lab took up projection VR to demonstrate semi-intelligent artificial agents with graphic bodies that use natural gestures. ALIVE also demonstrates advances in vision-based tracking systems.

The recent projection VR at the Electronic Visualization Lab at the University of Illinois at Chicago and at the MIT Media Lab bodes well for the alternative to HMD immersion. So does the success of the Mandala System in sports and entertainment. Still, the HMD type remains the more widely known brand of VR. But neither success nor widespread acceptance reveals the fundamental philosophical differences that splits VR development. Nor does widespread acceptance tell us how these two might complement one another. The relationship between the two types of immersion does not lend itself to quick and easy comprehension. But I think something important hides behind this difference, something which could lead to a deeper understanding of the long-range impact of VR on the human species.

14.3 VR Projection Images Do Not Re-Present, They Tele-Present

We have always been able to immerse ourselves in the worlds of novels, symphonies, and films, but VR allows us to move about and physically interact with artificial images. This feature is peculiar to VR and it is the field on which the major debate takes place.

This immersive feature has broad implications ontologically. First, virtual entities are not representations. They do not "present again" something that is already present somewhere else. Even telepresence robotics brings about a transformation of the remote entity, in which its properties become open in new ways; the doctor reconstitutes the patient and creates a new doctor-patient relationship through telepresence surgery. Virtual images are not like the images in paintings which we can in some sense "take" to be another outside entity which the graphic images represent. In VR, the images are the realities. As in the medieval theory of transubstantiation, the symbol becomes the reality. This is the meaning of telepresence. Telepresence or cyberspace is where primary entities are transported and transfigured into cyber entities. As another layer of reality, cyberspace is where the transported entities actually meet. They are present to one another, even though their primary bodies exist at a distance (the Greek teles). When the user is immersed in a virtual world, those entities encountered are real to the user – given the background of cyberspace. The user is immersed and interacts with virtual realities.

Granted that the immersion is part of VR, the question still remains: How are users best immersed in virtual environments? Should users feel totally immersed? That is, should users forget where they are (in a graphics environment) and see, hear, touch the world much in the same way we experience the primary phenomenological world? (We cannot see our own heads in the phenomenological world.) Or should users be allowed to see themselves as a cyberbody? Should they be aware of their primary body as a separate entity outside the graphic environment? What makes full-body immersion? The two different answers to this question split into the two meanings of full-body immersion. One goes into Myron Krueger's Videoplace and the other into the head-mounted displays of Tom Furness, Fred Brooks, and Jaron Lanier.

14.3.1 Tunnel Graphics and Boomerang Graphics: Perception and Apperception

Philosophically, the difference is profound. The HMD brand of VR produces what I call tunnel graphics or perception graphics. The projection brand of VR, on the contrary, produces boomerang graphics or apperception graphics.

Let me explain what I mean by perception and apperception graphics, so that I can then distinguish VR as perceptive and apperceptive VR.

The term "apperception" came into vogue with the Kantian philosophy of the 18th century. Immanuel Kant distinguished between perception and apperception[3]. Perception goes toward entities and registers the color, shape, quantity, and other properties of entities. Percepts have sensory qualities, whether visual, auditory, olfactory, or tactile. Apperception, on the other hand, sees not only entities but also that which accompanies the perception of any entities. With perception we see something. With apperception we notice that we are seeing something. Apperception implies a reflectedness, a self-awareness of what we are perceiving. For Kant, this aspect of perception meant that human beings enjoy a freedom and self-determination in their sensory activity that animals do not. Kant also believed that apperception makes possible a critical attitude toward what we perceive. Once we sense our separation from a stimulus, we can then enjoy the option of responding in various ways to the stimulus, perhaps even choosing to not respond at all to it.

In borrowing the term "apperception" from Kant, I use it not to name a general feature of human psychology but rather to highlight the advantage of one VR interface over another. In projection VR, the user experiences more than perception of entities. The user enjoys an apperceptive experience. The user's body enjoys the immersion experience without having to adapt to the system's peripherals (heavy helmet, tight data glove, calibrated earphones). The immersive experience here does not constrict but rather enhances the user's body. In turn, the projected graphics take a different phenomenological direction than perception-oriented systems. With perception-oriented graphics, the user's body gives way to the priority of the cyberbody and the tunnel-like perception of the graphics generated in real-time by the computer. That is why I describe the HMD graphic environment as tunnel vision. The user undergoes a high-powered interiorization of the virtual environment with a concomitant loss of self-awareness. (Discomfort detracts from an optimal and fully present self-awareness.)

Many computer graphics are perceptive graphics. They show us some things we can then constitute with our imaginations. Typically, graphics render representations of entities. Phenomenologically, VR graphics render entities directly. We see not only what the graphics refer to, but we identify with them. In VR, we see virtual entities. Graphics refers us to things. But HMD VR directs us exclusively toward the entities, into a tunnel-like perceptual field in which we encounter the graphic entities.

Apperceptive graphics, on the other hand, make us feel ourselves perceiving the graphic entities. Our freedom of bodily movement ranks in awareness alongside computer-generated entities. Apperceptive VR directs us towards the experience of sensing the virtual world rather than toward the entities themselves. To put it simply, HMD VR creates tunnel graphics, while apperceptive VR creates a boomerang telepresence that enables us to go out and identify with our cyberbody and the virtual entities it encounters and then return to our kinesthetic and kinetic primary body, and then go out again to the cyberbody and then return to our primary body, all in a deepening reiteration. Instead of Tron, we have Mandala. Not by accident was the first commercial projection VR named Mandala. The mandalas of Oriental art create an oscillation between outer perception and inner self-awareness. Unlike the hero of the film "Tron," we will not entirely lose ourselves in Mandala graphics.

HMD systems allow us to go "through the window" and engage

computerized entities, but apperceptive systems allow us to go further. We can advance both by entering cyberspace and at the same time celebrating the free play of our physical bodies.

The difference between perception and apperception VR systems means more than an ergonomic difference. Users often appreciate the freedom of movement possible with unencumbered VR, and the word "unencumbered" expresses that freedom. But "unencumbered" remains a merely negative definition, telling us only what this type of interface is not. By apperceptive VR, I suggest a positive definition of one of the crossroads facing VR development.

From the viewpoint of user phenomenology, the difference is one of a focused versus an expansive self. When we are not strapped into a helmet and datasuit, we can move about freely. The freedom of movement goes beyond being unshackled. It also means our spontaneity becomes engaged. Just watch for a few minutes the users of projection VR, how they turn and bend and use their bodies. Then contrast this with HMD users. The difference lies not in the software or the environment rendered. The difference lies in the hardware-to-human interface.

The creators of EVL's CAVE implicitly grasped this. By referring to Plato's Cave – as Thomas DeFanti fondly and frequently does –, the inventors at EVL recognize the human issue. The human issue is the issue of freedom embedded in the hardware-to-human interface of VR. Around 425 B.C., the philosopher Plato wrote, in Section VII of *The Republic*, a story he thought worthy of his teacher Socrates. The story tells of a cave and its age-old debate about the status of symbols, images, and representations. In Plato's parable, the people chained to the floor of the cave enjoy no physical mobility, and their immobilized position helps induce the trance that holds them fixed in its spell. They see shadows cast by artificial creatures ("puppets") held up behind them. The puppets have been created by human beings who want the prisoners to accept the shadows as the only real entities. The cave consequently becomes a prison rather than an environment for spontaneous behavior. Do not confuse Plato's cave with the caves of Lascaux. Plato's cave is a dungeon. Similarly, we may find that HMD-type VR can facilitate a higher level of human productivity and an information-rich efficiency – whether the application be flying aircraft, doing training, or holding meetings in a corporate virtual workspace, but it does so by exacting a human price.

Plato's story ends by having one of the prisoners escape. The prisoner escapes the dark dungeon and sees the primary light of sunshine and the experience of true entities. For Plato, the sunshine was the sphere of individual thinking and mental ideas. As long as the person stays fixed in a purely receptive mode, the mind lives in the dark. By climbing out of the cave into the sun of well-thought ideas, the prisoner ascends to the primary and true vision of things. With VR, we must reverse the metaphor. To escape from tunnel VR, we must re-discover the primary world, and this world is the primary body that already exists outside electronic systems. Our liberation is to enhance and deepen our awareness of the primary body.

From the perspective of user somatics, the difference between apperceptive and perceptive VR is one of the primary body versus the cyberself construct. The term "somatics" derives from Thomas Hanna[4]. Hanna defined somatics as the first-person experience of one's own body – as opposed to third-person accounts of one's body from a scientific or medical point of view. Somatic awareness is the line where conscious awareness crosses over into the autonomic nervous system, breathing, balance, and kinesthetic bodily feedback. The more we identify with a cyberself graphic construct, the less primary body somatics we preserve. Human attention is finite. When our attention becomes stretched and overextended, we feel under stress. The HMD tunnel may provide the greatest tool for training and for vicarious experience, but it exacts the greatest price on the primary body.

14.3.2 The Hope of a Complementary System

The construction of the self in cyberspace signals the highest risk in human evolution. If every technology extends our senses and our physical reach, then virtual reality extends us to the maximum because it transports our very selves into the electronic environment. If our contemporary culture already tends to stress and overextend our finite attention, then the VR tunnel holds great dangers. We may lose our way in the tunnel. We may lose part of our selves, our health, our body-mind integration. With this fear in mind, I discussed in my book, *The Metaphysics of Virtual Reality*, the allied phenomena of AWS (Alternate World Syndrome) and AWD (Alternate World Disorder).

What I want to do here for the first time is suggest a way we could capitalize on the fundamental philosophical differences between projection VR and HMD VR. Right now, the two approaches have opposing camps in the VR industry. In the future, the two camps might coalesce into a more effective and healthful holistic system. Let me describe one possible scenario in which the two VRs might merge.

The reference point for the merger comes from non-Western culture. From an evolutionary standpoint, our Western technological systems may eventually have recourse to cultures outside themselves in order to resolve dilemmas and internal paradoxes. The reference I will use is the ancient art of T'ai Chi Chuan, a Chinese variation of Yoga that arose in a martial arts context hundreds of years ago. T'ai Chi is a series of body movements, often performed very slowly, designed to sew together attention and physical body so as to increase the chi or felt internal body energy. The cultivation of chi results in more vibrant health, flexible joints, and overall enhanced biofeedback. I use T'ai Chi for a reference point also because I am intimately familiar with it after years of study, practice, and teaching. While I cannot explain here all the details of the T'ai Chi connection, I have found many people in the computer industry who understand this reference enough to understand the direction I suggest, and some have even expressed a desire to take up the project I envision using T'ai Chi.

Here is the theory. Because time spent in HMD VR tends to constrain and tunnel human attention into perceptive graphics, and because every technological advance exacts a price or trade-off, we should allot a corresponding amount of time for projection VR. We should combine projection VR with HMD VR just as we combine a decompression chamber with scuba diving. Scuba divers check time tables to find a ratio between time spent undersea and time needed in a decompression chamber. They then spend a certain amount of time in the decompression chamber so their deep-sea diving will not cause the scuba diver to suffer internal injury. Similarly with VR. The VR user should have a corresponding decompression procedure after spending a couple hours in HMD VR. The projection VR provides an analogous decompression in that projection VR utilizes technology to make a smooth transition from cyberbody to primary body. Rather than an abrupt shock between cyber and primary worlds, the user brings attention back into the primary mind-body and re-integrates the human self.

Here is the scenario. Projection VR provides a T'ai Chi master generated by computer graphics. The graphics are based on movements and postures modeled by actual T'ai Chi masters. The computer-generated master can teach not only the series of movements, but can also lead Chi Kung exercises, do push hands, and even enter into sparring bouts with the user. This VR decompression chamber will link the user to the primary world smoothly and with an intensity that reclaims the integrity of conscious life in a biological body. Such a procedure will help offset the disintegrating aspects of reality lag. The VR experience can grow into a health-enhancing rather than health-compromising experience.

Using this combination of apperceptive and perceptive graphics, we could bring about a balance. Both types of graphics can enhance each other. VR could thereby produce some intensely focused virtual environments that lead to more alert and self-aware human beings.

14.4 References

[1] I first described the seven meanings of VR in Chapter 8 of my book, Heim, M. (1993). *The Metaphysics of Virtual Reality.* New York: Oxford University Press.

[2] Boorstein, D. J. (1978). *The Republic of Technology: Reflections on Our Future Community.* New York: Harper & Row, p. 7.

[3] Kant introduced the distinction throughout his *Critique of Pure Reason* of 1781.

[4] Hanna, T. (1988). *Somatics: Reawakening the Mind's Control of Movement, Flexibility, and Health.* Reading, Mass.: Addison-Wesley.

15

Audio Display from the Simple Beep to Sonification and Virtual Auditory Environments

Rory Stuart[1]

15.1 Introduction

In part because audio has been much under-utilized in human- computer interfaces, the reader may find this area less familiar than many of those discussed in this conference on Understanding Images. This overview of the use of audio displays in human- computer interfaces will present and discuss the types of information that audio has been used to convey, some background on pertinent findings of perceptual research, the work that has been done in designing and evaluating audio displays, and the issues that are raised by this work.

How can audio be used to convey information, and how does the user understand information presented sonically? Several approaches to the use of audio to convey information will be described. As with visual images, one brings to an understanding of audio one's history and cultural experiences, one's perceptual capabilities, and one's ability to learn new associations and meaning.

There are two types of audio that will not be in the primary focus of this discussion of audio display, but some comments on them are in order. The first type is speech (digitized or synthesized). The focus here will be primarily on non-speech audio, not on the understanding of speech. Humans develop a rich body of experience in communicating with speech during their lifetimes, and this is an advantage of using speech; one of the shortcomings of using speech is that its use results in language-dependent interfaces, which is a problem for

[1]Intelligent Interfaces Group/Artificial Intelligence Laboratory, NYNEX Science and Technology Inc., White Plains, NY 10604

users who do not speak the language. It is worth noting that there is an aspect of speech that conveys information beyond the content of the words and language used, and this is its prosody - - its emphasis, rhythm and melody. Synthesized speech in particular often fails to convey some of the information conveyed by a human speaker because of missing prosodic cues, though prosody can be incorporated in speech synthesis (as it has been by scientists at NYNEX – see e.g. [1]). Another element that may be conveyed by the acoustic properties of computer-presented speech if appropriate cues are synthesized is the location of its source [2]; in the discussion below of virtual auditory worlds, the reader should keep in mind that the sound sources presented may be speech as well as other audio.

The second use of sound that will not be a focus here is "incidental" audio: sound used to create mood, as in background music. Although the reader may sneer at the thought of "Elevator Music for Interfaces," there may be something to be learned from the cinematic use of music to enhance narrative and convey mood and rhythm. Like many of the conventions of film, this has not been effectively adopted in human-computer interfaces [3], and will not be discussed here. Instead, the focus will be on non-speech audio that conveys information.

The information that can be conveyed by non-speech audio ranges from simple binary state to complex multidimensional data. An example of conveying simple binary state information (common in current interfaces) is the case when the user is typing in a multi- window graphical user interface (GUI), and the cursor is in one of two states: located within a window, in which there is no audio, or not within a window, in which case the system beeps when the user enters a keystroke. An example of conveying complex multidimensional data is in the sonification displays described below. In general, the information displayed can include warnings and alarms, feedback on interaction, display of data, and information contributing to situational awareness.

Audio displays may be combined with visual displays of text and graphics, with a continuum from total redundancy to total lack of redundancy between the audio and visual information presented; and they may be presented without accompanying visual display (e.g. in phone-based interfaces or interfaces for visually impaired users).

Among the major issues that arise in audio displays is their interaction with other audio displays (e.g. in the case of multiple alarms) and with sounds in the user's environment; the choice of sounds and choice of mapping between the information to be displayed and its representation (with a range of approaches taken, as described below); and the evaluation of the applicability of audio display for different tasks (e.g. the use of virtual acoustic display in aircraft cockpits or in teleoperation).

After briefly reviewing motivations for the use of sonic displays in human-computer interfaces, and summarizing some relevant background in auditory perception, we will turn to an examination of different applications and approaches to audio display and the issues associated with these approaches.

15.2 Motivations for the Use of Audio

Most of the widely used computer-human interfaces may be thought of as "impoverished" environments, in the sense that they do not use many of our perceptual and motor capabilities. Typically, they do not take advantage of our 3-D visual capabilites, peripheral vision, tactile and olfactory sensitivity, sense of equilibrium, kinesthetic sense, proprioception, motor capabilities (except for eye and finger movements), or production of speech [4]. In the audio realm, they often provide only a very limited repertoire of intentional sounds (e.g. an error beep), although the user may get information about the status of the system from sounds that are not planned for by the designers, such as the sound of the hard drive spinning. But these interfaces do not take advantage of our capabilities of auditory localization, or of monitoring and distinguishing multiple simultaneous streams of sound. There are several approaches to enriching these "impoverished" human-computer interfaces and expanding the bandwidth in human-computer interaction that are being explored, and among them are increased use and new uses of audio.

Why use sound at all? For eyes-busy tasks, sound can be used to command attention and alert the user. The simple beep, which is the only nearly ubiquitous use of sound on personal computers to date, conveys little information, but is useful for what it does convey. Wrapped in thought and writing text or code, one could easily miss the cursor moving out of a window on the computer screen (as one's optical mouse slides an inch or two), were it not for the beep. For visually impaired users, there may be no better alternatives to the use of sound. The ears can take in information from all directions (whereas the eyes can only take in information from their point of regard and peripheral field), and can take this information in simultaneously. For users without visual impairment, the ears can direct the eyes. Sound can provide situational awareness and permit the monitoring of background signals. Finally, the available interface platform may dictate that there are no viable alternatives to audio, for example for a phone-based interface (PBI). Although PBIs often use spoken prompts, our research has shown that, at least in some contexts, non-speech audio can be preferable. At the top level of a voice activated dialing interface, we wanted a minimal prompt to maximize the speed and transparency of calling, yet found that an abbreviated voice prompt was often interpreted differently at different times by the same user. In contrast, users were more likely to treat a tone prompt as having a unique and unambiguous meaning [5].

For applications in which the target users are a diverse group who speak different languages, the use of non-speech audio, like the use of well-constructed graphical traffic signs, offers the potential of language-independent interfaces.

15.3 Auditory Perception

In a broad sense, the issues in audio display may be seen as falling into three categories: perceptual issues, display issues, and design approaches. The latter two will be discussed in the context of specific systems and applications. It is worth first noting some points about auditory perception that are salient to all of the work that will be subsequently described. These include auditory scene analysis and stream segregation, detectability and masking, interactions between sounds, auditory fatigue, and the psychology of everyday listening. In the case of virtual acoustic displays, they also include auditory localization and intersensory interactions. Among the obvious properties of sounds and our perception of them are that sounds necessarily have a temporal component; unlike vision, which requires that our eyes be directed towards something for us to see it, we perceive sounds coming from sources in all directions; and sounds that are fairly constant tend to stop being consciously noticed but changes in them rapidly draw our attention (as when there is a sudden change in the sound of a car engine, or the sound of a central air conditioner suddenly stops).

15.4 Auditory Scene Analysis

Bregman [6] provides an excellent description of current knowledge in the area of auditory scene analysis. There are typically many sources of sound in our environment within our hearing range, and the pattern of acoustic energy that reaches our ears is a mixture of these. The goal of scene analysis in general is to recover separate descriptions of separate objects that occur in the environment. Auditory scene analysis is the process by which the human listener perceptually groups acoustic events within a complex auditory environment into discrete "objects," and is able to build mental descriptions of these separate sound sources or events. As a simple example, and one that suggests how early this process is in place, Bregman points out that even a baby that starts to imitate her mother's voice does not insert imitations of the squeaks of her cradle that were happening simultaneously, and has therefore been able to distinguish the squeaks as not being part of the perceptual "object" formed by her mother's voice.

The study of scene analysis has primarily been conducted in the domain of vision, and has been a focus of Gestalt psychologists. In vision, we use some of the light that is reflected from objects, bounces around between the objects, and reaches our eyes, to form descriptions of the individual objects in the environment. We visually group things according to well-described principles, and we "fill in" the shapes of objects partially occluded by other objects, and do so in predictable ways. Auditory scene analysis has been less studied than visual scene analysis, but Bregman presents a theoretical framework and the processes that comprise its components.

The grouping of sequential sounds (e.g. a sequence of tones) into a single perceived event is termed "auditory stream segregation," and is influenced by the rate at which the sounds occur, the difference between their frequencies, and their timbres, as well as the location of their apparent sources. "Spectral integration" is the process through which we group different spectral components of complex sounds heard simultaneously, and associate them with their sources;

it involves comparison through time between a current spectrum and that which preceded it. These, and other of what Bregman terms the "simple" processes, rely on relatively constant properties of the everyday world, such as the tendencies of sounds to be continuous, have components that begin and end together, and move through space along paths at fairly smooth rates. In addition, he postulates schema-based stream segregation, which involves our using mental constructs formed through our knowledge and experiences with certain types of familiar sounds (such as common environmental sounds, speech, and music) to group current sounds to which we are exposed.

15.5 Detectability and Masking

Detectability is particularly critical in the case of alarms, and will be discussed more in that context. Detectability of an individual sound in a controlled setting is simple to determine, and predictable from our knowledge of psychoacoustics. But, in real-world audio displays, things become more complex, with the possibility of multiple simultaneous sounds and sound sources. Masking describes the phenomenon where one sound "masks" or obscures some of the spectral components of another sound. In fact, sounds can be masked by other sounds that immediately precede or follow them, so this should also be considered. Masking depends on loudness (with louder sounds generally masking softer ones) and frequency (with lower frequencies generally masking higher ones). Masking especially tends to happen between sounds of frequencies within approximately one-third of an octave range (the "critical band"). If the sounds that will be displayed and other sounds from the user's environment are known, accurate predictions of masking are possible. There has been work done on reducing masking in virtual auditory worlds (referred to in a subsequent section), but in the case of non-spatially presented audio, it is important to consider the combinations of sounds that may be presented simultaneously or in close temporal proximity, and the sounds from other sources that may be in the user's environment (e.g. in the cockpit of an aircraft) in order to evaluate masking and its effects on detectability. For more on masking, as well as psychoacoustical information regarding pitch, loudness, and duration, and their implications for audio displays, see Buxton et al. [7].

15.6 The Psychology of Everyday Listening

Much of the research on auditory perception and psychoacoustics has examined what might be thought of as "musical listening." This embodies things like discriminability of pitches and melodies. But do we actually listen to events in the world in the same way we listen to music? Bill Gaver of Rank Xerox Cambridge EuroPARC discusses the idea that there is a psychology of everyday listening, which may be contrasted to that of musical listening. In musical listening, the attributes of the sound itself (pitch, timbre, and loudness, and, on a larger scale, melody, harmony, rhythm, etc..) are the focus of attention; in everyday listening, the attributes of the source of the sound are the focus of attention. For example, Gaver would maintain that, if we hear a rock hit a can, we don't perceive this in the way we would a musical event (such as attending to its pitch), but instead perceive things about the sources (such as how large and how filled the can is, and how hard the rock was thrown). An anecdote I have heard about a jazz musician who had such good "ears" that he immediately exclaimed "C#" when the shoe of his colleague squeaked highlights the fact that this is not how most of us listen to the squeaking of shoes or other everyday sound events.

Gaver describes a hierarchy of attributes of the sources of everyday sounds and their effects on the soundwaves produced. Object properties have characteristic effects, e.g. size is associated with frequency and bandwidth, force is associated with amplitude, and so forth. The taxonomy also describes aerodynamic, liquid, and vibrating solid sounds, where, for example, liquid sounds include dripping and splashing, and dripping sounds have attributes associated the properties of the liquid (such as its viscosity and density) and impacting object (such as its size, shape, and mass).

Gaver and others have used the characteristics of everyday listening in designing audio displays for human-computer interfaces, as will be described below.

15.7 Approaches and Applications of Audio Displays

We now turn to approaches and applications of audio displays. These include warnings and alarms, interaction feedback and status information, displays of abstract data, and displays to enhance situational awareness. Some of these approaches have had widespread use in a variety of systems (e.g. warnings and alarms), and others have only gotten to the prototyping stage. But most suggest further potential applications of auditory display in human- computer interfaces, and all reveal human factors and display issues regarding the perception and understanding of the displayed information by the user.

15.7.1 Warnings and Alarms

Warnings and Alarms are perhaps the most conventional and certainly the most widespread use of audio in human-computer interfaces. The simple beep that warns the user of an error has become common across most computer platforms (and in the case of many personal computers, is the only use of audio). Monitoring systems often have non-speech alarms as well as speech alarms. Audio can be especially good at attracting attention, and attention to problems indicated by alarms may be life-critical (e.g. on an airplane or at a nuclear power facility). There is considerable experience with warnings and alarms to be drawn from domains other than human-computer interaction (e.g. with ambulance sirens and foghorns), although complex computer-based systems may produce alarm situations that would seldom be found in other domains. Sanders and McCormick [8] describe the more than 60 alarms and audio warnings that occurred simultaneously during the crisis at Three Mile Island. Unfortunately, audio alarms and warnings have often been created individually without considering their interactions, and in situations of great complexity like the Three Mile Island crisis, this severely limits the usefulness of the alarms – they may mask each other, be difficult to distinguish, and add to stress and confusion.

Audio warnings and alarms must also be considered in terms of the tasks and environment for which an interface is used if they are to be designed effectively. An alarm, for example, may be made "too alarming" – if the first activity triggered by an alarm at a nuclear power plant, factory floor, or airline cockpit is to try to silence the alarm (e.g. because it is so jarring and loud that people can not hear each other over it in order to communicate and solve the crisis), valuable, even life-critical, time may be wasted. Dr. Roy Patterson of the Applied Psychology Unit of the UK Medical Research Council [9] has championed alarm systems that are as unobtrusive as possible, while still commanding the user's attention.

15.7.2 Status Information – Auditory Icons: Gaver

Gaver [10], as noted above, is an exponent of using our capabilities at "everyday listening" in human-computer interaction, and uses the term "auditory icons" to describe the use of sounds to represent information about events in the computer in a way that will convey meaning to the user by analogy with everyday events. He has created an interface that uses audio icons to represent Apple Macintosh Finder events such as opening files, dragging folders, scrolling windows, and emptying the trash can [11]; and an interface that uses auditory icons to display information about a simulation of a soft-drink bottling plant [12]. There are other presentations of simple status information that are not uncommon, but they generally require learning arbitrary associations between sound (e.g. a sequence of beeps) and system status, where Gaver's auditory icons let the user map knowledge about everyday sounds to the event at hand (e.g. the sound of the file being dragged gives a clear indication that it is large, and there is no doubt from the sound displayed that it has been dropped in the "trash").

Another example of the use of audio to display status information is in systems to aid visually impaired users with GUI-type interfaces. One such system is described by Edwards [13].

15.8 Displaying Abstract Data: Sonification and Auralization

In displaying data, the appropriate method of presentation for some types of data, such as seismic activity, may be auralization, which is the direct presentation of the data (e.g. seismic "rumble"), altering only its time scale or modulating its frequency to make it more easy to perceive. In this case the main issue to be addressed is how to alter frequency and time scale to enable the user to effectively detect patterns and anomalies. The best available guidelines for auralization are to use knowledge of auditory perception (e.g. the range of frequencies for which discrimination is most accurate), combined with some iterative evaluation with target users. For most data, however, it is also necessary to choose some mapping of the data to the audio display – the data (such as economic indicators) itself does not already have a sound as does seismic activity. This mapping may be anything from a direct mapping between numerical values and the values of sound parameters (e.g. between temperature and frequency of a tone), to a more semantic-level mapping (e.g. between selected characteristics of the data and melodic motives). Sonification is the term used to describe the presentation of data via some mapping of data parameters to sound parameters (for more on the current state of the art in sonification, see Kramer [14, 15, 16]).

15.8.1 Sonification: Bly

In Bly's [17] dissertation work at the University of California at Davis, she ran eight experiments in which a variety of data sets were presented in different ways through sound. For representing multivariate data, she directly mapped data parameters into seven sound parameters: pitch, volume, duration, fundamental waveshape, attack envelope, and additions of 5th and 9th harmonics, and found that auditory representation of the data improved subjects' ability to distinguish the datasets. She also represented time-varying data, assigning waveform to individual functions (which were displayed simultaneously) and pitch and volume to parameters within a given function; and represented logarithmic data, using only pitch. In all cases, she found that the auditory representation of data conveyed useful information, both by itself and in combination with graphics [18]. Although she has said that, knowing what she now knows, she would choose a different mapping between data and sound parameters, her empirical testing with subjects provided an early demonstration of the value of sonification.

15.8.2 Sonification of Chemical Spectra: Lunney & Morrison

Lunney & Morrison at East Carolina University have worked since the early 1980s on aids for the visually impaired scientist and have especially focussed on auditory presentation of chemical data such as infrared spectra to visually impaired chemistry students [19, 20, 21, 22]. Their approach, another early example of sonification, involves mapping the major features of a chemical spectrum to musical notes of the chromatic scale (within an eight octave range). They convert the continuous infrared spectrum to a stick spectrum and map those peaks to chromatic notes, with duration of each note, in their first type of display, corresponding to intensity of its infrared peak; discrete rather than continuous values of both pitch and duration are used (quantization, i.e. chromatic notes vs. pure frequency scaling, and divisions into whole-notes, half-notes, quarter-notes, and so on, rather than pure temporal scaling). Their auditory display presents the information in three forms: 1) notes played sequentially in order of descending pitch (with duration, as noted, corresponding to intensity); 2) notes played sequentially with equal duration, in order of decreasing peak intensity; and 3) notes played simultaneously in a chord, with all notes played for equal duration. As they do not use sound characteristics other than frequency and duration, many of these other characteristics can be chosen by the user (e.g. what waveform to use in presenting the data). They have noted a high success rate by their users in distinguishing chemicals by this method, but this has been the product of informal tests, and no reports of rigorous formal tests have yet been published.

15.8.3 Sonification of Multivariate Time Series Data: Mezrich, Frysinger, & Slivjanovski

Mezrich, Frysinger, & Slivjanovski [23] have applied auditory displays to exploratory data analysis of multidimensional time series data. Here, the challenge for the user is to find patterns in complex data that could not have been predicted. Unlike Lunney & Morrison, their target users were not visually impaired, so they designed displays that would present data graphically as well as sonically (and their displays were redundant, i.e. the same information was displayed both graphically and sonically). Also unlike Lunney & Morrison, their focus on data with a temporal component attracted them to the idea of letting the user "play" the data, in the sense that we can "play" a movie. One domain from which they chose data to display is that of economic indicators, and in their paper in the Journal of the American Statistical Association published in 1984, they describe a display of the components of the Data Resources Inc. (DRI) Economic Boom Monitor. This monitor is a composite of eight indices (normalized and weighted) that describe the cyclical state of the economy. Mezrich, et al. look to their approach not to replace graphs in displaying data, but to provide another display approach that will provide new insights. For example, a display such as the Boom Monitor must necessarily hide the relationships between its sources of data, and Mezrich, et al. propose their approach as one that will facilitate the exploration of these relationships

Like Lunney & Morrison, Mezrich, Frysinger, & Slivjanovski performed quantization rather than pure frequency scaling; unlike Lunney & Morrison, they chose to map data to notes of the A major scale, rather than to the chromatic scale. Mezrich et al. later reported applying their display approach to sonification of DNA sequencing.

15.8.4 Exploratory Visualization Project: Smith, Bergeron, & Grinstein

Smith, Bergeron, & Grinstein [24], working on the Exvis project at the University of Lowell, have developed a visualization facility that graphically and sonically displays information which is high dimensional but not necessarily time-related. The user triggers the auditory data presentation by using a mouse to move a cursor over the visual display. Each data point in the visual display is represented by what the developers call an "icon" (note that this is distinct from the "auditory icons" of Gaver)[10], with the attributes of the icon determined by the data it represents; and when the cursor passes over each icon, it generates a sound. The sounds produced, i.e. the mapping between data and sound, can be chosen by the user. Smith et al. produce in this display what they describe as textures – both the visual textures produced by many graphical icons, and the analogous auditory textures; their displays sound more "textural" and less "musical" than those of Lunney et al. or Mezrich et al. After initially presenting the sound display monaurally, they developed the system to present sound stereophonically through a pair of speakers, with sounds displayable along the horizontal plane and with some effect of depth through the use of reverberation. One of the user's choices is whether the horizontal stereo position of the sound should correspond to its position in the visual display, though their system does not permit true two-dimensional display because they do not have an available technique for real time synthesis of elevation cues over loudspeakers.

15.8.5 EARCONS: BLATTNER, SUMIKAWA, & GREENBERG

Blattner Sumikawa, & Greenberg [25] introduce the concept of earcons, which they consider to be the "aural counterpart of icons." Earcons present non-speech audio messages that give the user of a human-computer interface information about computer objects (e.g. files, menus, or prompts), operations (e.g. editing, compiling, or executing), or interactions between objects and operations (e.g. editing a file). They use the principles of musical listening, presenting motives (melodic phrases, i.e. pitches presented in a rhythmical sequence) made up of notes in the major and minor scales. Unlike in sonification, where data is mapped to sound parameters and the listener attends and attempts to find meaningful patterns, Blattner et al. use sound to represent concepts that the designer already has in mind and intends to express through sound. Rhythm, melody, timbre, register, and dynamics are used to encode information. A particular rhythm, for example, may indicate an "error," and the type of error may be indicated by the choice of pitches or timbre in a motive that uses that rhythm.

The user must learn the mapping intended between motive and meaning, and the designer of earcons attempts to make design choices that will permit the intended information to be communicated while minimizing the human memory required to retain an earcon. Blattner, et al. categorize the methods of earcon design as 1) representational earcons (not compounded), 2) combined earcons, and 3) family earcons (hierarchical and transformational, e.g. inheriting attributes). Although Blattner et al. presented system status information and interaction feedback with their earcons, the principles embodied in their work could be used in other applications, where intelligence in the system would produce the more "semantic" mappings that they use.

15.9 Virtual Auditory Worlds

Virtual auditory worlds (VAWs) present audio displays so that the user perceives himself or herself to be "in" the sonic environment. The user, generally wearing headphones, perceives the displayed sounds as coming from positions or trajectories in the surrounding world, which are maintained regardless of the user's motion. VAWs use position tracking and digital signal processing to synthesize the cues humans use to localize sounds, and have only been created during the past few years. The study of these auditory localization cues is ongoing. Although human sound localization has been studied since at least as long ago as the late 18th century by Venturi, it is still not entirely understood. Approximately two centuries ago, Venturi designed an experiment in which he circled subjects in an open field at a distance of forty meters, playing occasional notes on his flute; he concluded that the difference in volume of sound reaching the two ears (interaural intensity difference) was a cue for localization. Later, interaural time and phase differences were also identified as cues for localization. The importance of pinna filtering has only been recognized during the past thirty years (e.g. see [26]), and an example of a localization phenomenon that is still not entirely understood is that of "externalization," or hearing things as coming from outside of the head [27]. Many studies have focused on a single stationary sound source presenting a single sound (single frequency or white noise) in anechoic conditions to a stationary listener. In these conditions, diotic stimuli (same in both ears) and dichotic stimuli (different to each ear) have been used to study the four localization cues mentioned: interaural time, intensity, and phase differences, and pinna filtering.

In pinna filtering, the pinna (or external ear) essentially acts as a filter that distorts incoming sound signals depending on their direction and distance in a way that permits the listener to decipher the direction/distance information. Blauert [28] gives an excellent presentation on this pinna-related transfer function, and notes that the acoustical effect of the pinna is based on reflection, shadowing, dispersion, diffraction, interference, and resonance. As the entire head has effects on the incoming sound signal, this direction-dependent filtering is called the head related transfer function (HRTF). If the head position and orientation of users are tracked, digital signal processing can be used to synthesize HRTFs in a real-time system and, playing these processed sounds over headphones to the user, a VAW is displayed. Two manufacturers (Crystal River Engineering and Bo Gehring) currently offer commercially available real-time systems to synthesize HRTFs.

As noted above, much of the work in this area has addressed single stationary sound sources in anechoic conditions presented to a stationary listener. But in our everyday listening experience, things are much more complex. There is reverberation as sounds bounce off many sources, and this may help us identify characterisics of the room or acoustic environment we are in, as well as aid in identifying our distance from the sound source. Mershon & King [29] identify reverberation as one of the few cues for distance, and perhaps the best cue for absolute distance, but it is thought to make it somewhat more difficult to identify the direction of a sound source. In our everyday listening experience, there are multiple simultaneous sounds and sound sources, which introduce factors such as masking (where one sound "masks" or obscures some of the spectral components of another sound). Doll, Hanna, & Russotti [30] suggest an approach for minimizing masking effects in VAWs by maintaining adequate angular separation between sound sources and keeping background sound sources uncorrelated. In the listening conditions to which we are accustomed, sound sources can move (or we can move), which may introduce doppler effects (which, for example, help us determine when an ambulance siren is approaching us by shifts in its pitch), and Grantham [31], Strybel [32], and Strybel, Manligas, and Perrott [33] have studied the human capability to auditorily detect the motion of sound sources. Finally, in our everyday listening experience, there are cues in addition to audio which produce intersensory interactions. These, for example, account for the ventriloquist effect, in which there is a conflict between visual and auditory cues and the illusion is created that the ventriloquist is "throwing" his or her voice, and producing it from the moving mouth of a dummy. Ventriloquism is an illustration of the finding that vision has a powerful dominating effect on the localization of sound [34].

Under complex real- world listening conditions, with multiple sound sources bouncing off different surfaces, and following multiple paths; the listener has to do auditory scene analysis as described above, perceptually integrating signals arriving at different times and distinguishing individual sound sources. In the most advanced commerically available virtual acoustic display system (Crystal River's highest end system), anechoic cues for sixteen sound sources can be given, or four sound sources can be displayed incorporating simulation of first-order reflective environments, with user control of the location and absorption characteristics of reflective elements (such as virtual walls and floor) [35]. This impressive system, which requires gigaflop computing, still presents VAWs that fall far short of the complexity of most everyday non-synthetic environments.

VAWs can be used to broaden the human computer interation bandwidth by utilizing more of the human perceptual and motor capabilities [36] and specifically to 1) give the user an interface that provides awareness in all directions (including those directions that are outside of the field of view) and directs the user's eyes (cf. [37, 38, 39]); 2) provide situational awareness of environmental surroundings and the orientation of the user (cf. [40]); 3) help the user separate multiple streams of sound and direct attention selectively by, for example, using the "Cocktail Party Effect" [41] to attend to a single speaker; and 4) convey information synesthetically, e.g. to convey normally-visual orientation cues to a visually impaired user [42], or to convey touch information if there is no force/tactile feedback available in the interface. Because the technology to create VAWs has been available for such a short time, many of their applications are still only in the discussion or research phases, but these include enhanced displays for air traffic controllers, being researched by NASA Ames in collaboration with the Federal Aviation Administration [43]; target acquisition aids, presumably for military purposes, being studied by several researchers (e.g., [39, 44]); enhanced sonar displays, studied by the Naval Submarine Medical Research Laboratory [38] and by the Naval Command Control and Ocean Surveillance Center [45]; and aids to the visually impaired [46].

15.9.1 Wenzel: Implications for a Virtual Acoustic Display of Individual Differences in HRTFs

Elizabeth Wenzel leads a team at NASA Ames that has done some of the leading work on virtual acoustic displays. HRTFs were measured by Wightman and Kistler at the University of Wisconsin at Madison using probe microphones placed in the ear canals of subjects and playing sounds from different locations in an anechoic chamber [47]. Using these HRTFs and working with Scott Foster of Crystal River Engineering, a virtual acoustic display system (mentioned above) was developed. Working with Wightman and Kistler, Wenzel then studied how well subjects could localize sounds using an HRTF that is not their own. The process through which Wightman & Kistler measure HRTFs is rather elaborate, and it would be impractical to provide VAWs to users on a wide scale if every user had to go through this process and their individual HRTF synthesized in their virtual acoustic display. Wenzel et al. found that, using the HRTF of a person who was rather good at localizing real-world sounds, most subjects did well at localizing sounds presented via a virtual acoustic display, although certain characteristic errors are fairly common [48]. The work of Wenzel and her team suggest the viability of more widespread use of VAWs which use generic HRTFs.

15.10 Future Research

The field of audio display is ripe with areas for future research. Among these are enhanced localization cues, automatic Foley generation, the integration of sonification, virtual acoustic display, and telepresence audio, and the implementation tools to better support the spectrum of audio displays described.

15.10.1 Enhanced Localization Cues

The work of Wenzel et al. suggest that people can adapt to HRTFs that are not their own. This suggests the intriguing possibility of synthesizing enhanced localization cues, which allow the user to become a "super-localizer." Stuart [49] gives the example of the barn owl which, with its assymetric ears, can pinpoint the location of a distant prey by sound alone, but questions whether a human could adapt to VAW localization cues enhanced like those of the barn owl. Durlach [50] proposes to enhance localization cues not to mimick the capabilities of other species, but to compensate for shortcomings in the human localization system by, for example, uniformly exaggerating spectral and interaural differences as though our head and pinnae were enlarged. Adaptation to these altered sensory cues is still an area for future research.

15.10.2 Representational Sounds: Automatic Foley Generation

Sounds displayed in virtual environments to date are often "representational" (in the sense that a painting is representational rather than abstract), i.e. like sounds that might be heard in the non-synthetic world. Examples are the sounds of a monologue, musical instruments, a toy airplane flying around a simulated courtyard, or a crowd of people talking. The interactivity is typically rather limited in that the user may be able to trigger a small number of sounds or move in the virtual environment and listen to the sounds from different positions. Lenat & Guha [51] point out that realism in virtual environments will require that, if a user acts in an unexpected way, the objects in the world still behave appropriately. For example, a virtual pencil-like device should not only be able to write on a virtual writing pad, but, if hit unexpectedly against a glass, produce an appropriate ringing sound. (It should be noted that, in spite of the focus of the popular press, realism is not necessarily a goal in virtual environments for many applications.)

15.10.3 Transduced Telepresence Audio, Spatialized Sonification, and the Integration of Virtual Acoustic Display

Transduced Telepresence Audio, and Sonification Teleoperation systems can provide virtual reality-like displays to a human controlling a remotely located anthropomorphic robot. In this case, there is a tight sensory-motor loop that gives the user the experience of being present at the remote location and directly manipulating objects there. If properly located microphones are placed in the "ear canals" of the anthropomorphic robot, the user can localize sounds at the remote location as though present there. If the robot is sufficiently anthropomorphic, and the lag (i.e. delay between when the user moves, the robot moves, and the user receives remote transduced sensory data from the robot's new position) is not enough to be perceived, such a system does not require any special processing of the audio, and does not produce any special challenges to the listener in understanding the audio display – such a system is so veridical that the user hears what they would if they were at the remote site. Indeed, early work on VAWs was done with a single speaker and a mannequin head that had microphones in its ears and was mounted on servo-driven assemblies and motion-coupled to a human operator. This system, the Gehring Auditory Localizer, preceded systems that synthesize localization cues via digital signal processing, and is described in Calhoun, Valencia, and Furness [40].

Real teleoperation systems are often quite different from that just described. For example, a more problematic case of transduced telepresence audio is in minimally invasive telepresence surgery. Dr. Satava has spoken of the importance of audio in telepresent surgery, where the surgeon can hear remote sounds made by the small surgical instruments inside the patient's gut. Clearly, the sound transducers in such a system would have to be tiny microphones, positioned closely together, and this would be very dissimilar to the microphones in an anthropomorphic robot. Adaptation and localization using the binaural sound displayed might be difficult without processing of the audio (e.g. interaural time delay would be minimal, and there would not be pinna filtering as there is with a mannequin head or anthropomorphic robot); processing this audio to reintroduce those cues is a technically challenging area for future research.

There have not yet been empirical studies of spatialized sonification. Given what we know of auditory scene analysis and auditory perception, there are reasons to believe that presenting sonification via virtual acoustic display and perhaps using spatialization of the sounds to provide additional information might help the listener recognize patterns and anomalies in the data.

Finally, the integration of virtual acoustic display, transduced telepresence audio, and sonification offers exciting challenges for future research. In a recent grant proposal, Gregory Kramer and the author proposed an auditory interface for the remote control of unmanned ground vehicles that would incorporate these three elements. Sonification, displayed spatially, would be used to convey system status information, and would be integrated with transduced telepresence audio, which would convey real-world environmental sounds from the remote location. How would the operator of such a system do at maintaining situational awareness while understanding the abstract data displayed? To date, empirical studies have examined these three types of audio display individually but not in combination.

15.11 Implementation Issues

Although the focus here has been on how audio can be used in human-computer interfaces and how it is understood by the user, implementation issues have been stumbling blocks for many of the audio interfaces described. Early researchers on sonification, such as Bly or Lunney & Morrison, had to build special hardware or interfaces to prototype their ideas. More recently, devices that use the standard Musical Instrument Digital Interface (MIDI) protocol have become commercially available due to the use of these devices in music. However, even researchers doing more recent work, who have used these commercially available MIDI devices, have noted the limitations of the MIDI protocol and their need to use the "System Exclusive" capabilities of the devices, effectively bypassing the MIDI standard and requiring device-dependent programming [24]. Few devices and prototyping tools designed for sonification are available yet, and this hampers researchers in this area. Virtual acoustic displays are now commercially available as noted above, but the low end systems permit only anechoic localization cues of few sound sources, and even the expensive systems have the limitations previously noted. Better modeling of propagation, including propagation delay, spreading loss, absorption, reflection, diffraction, non-uniform radiation, and transmission loss, should permit more veridical VAWs. The availability of more powerful easy-to-use inexpensive tools for building audio interfaces will facilitate more prototyping and empirical evaluation, which should in turn lead both to better audio interfaces and wider deployment of these interfaces.

15.12 Conclusions

Although certain uses of audio in human-computer interfaces have been widespread – such as the simple beep, and warnings in aircraft cockpits – audio has generally been underutilized, and many of the display approaches described, though shown to be effective in lab studies, have not been commercially deployed. But the necessary understanding of auditory perception has progressed, and the range of efforts in auditory display described herein suggest the potential for more widespread use of audio to convey various types of information. It is hoped that this overview has helped point interested readers to pertinent references for further information and has suggested promising areas for future research; and that, perhaps, some readers will use this information to design human-computer interfaces that make effective use of audio.

15.13 References

[1] Silverman, K., Kalyanswamy, A., Silverman, J., Basson, S., & Yashchin, D. (1993). Synthesizer intelligibility in the context of

a name-and-address information service. *EUROSPEECH '93, Proceedings of the 3rd European Conference on Speech Communication and Technology.* Berlin, Germany: ESCA. 2169-2172.

[2] Begault, D.R., & Wenzel, E.M. (1993). Headphone localization of speech. *Human Factors* Vol. 35, No. 2. 361-376.

[3] McKendree, J, & Mateer, J.W. (1990). Film techniques applied to the design and use of intelligent systems. *MCC Technical Report Number ACT-HI*-260-89. Austin, TX: Microelectronics and Computer Technology Corporation.

[4] Stuart, R, & Thomas, J.C. (1991). The implications of education in cyberspace. *Multimedia Review* 2(2), (Summer), 17-27.

[5] Lawrence, D., & Stuart, R. (1990). Case study of a user interface for a voice activated dialing service. In D. Diaper, et al (Eds.), *Human-Computer Interaction – Proceedings of INTERACT '90.* Elsevier Science Publishers B.V., North Holland, pp. 773-777.

[6] Bregman, A.S. (1990). *Auditory Scene Analysis.* Cambridge, MA: MIT Press.

[7] Buxton, W., Gaver, W., and Bly, S. (1989). The Use of Non-Speech Audio at the Interface. *Tutorial #10, CHI '89*, New York: ACM.

[8] Sanders, M.S. & McCormick, E.J. (1987). *Human Factors in Engineering and Design.* New York: McGraw-Hill.

[9] Patterson, R.R. (1982). Guidelines for auditory warning systems on civil aircraft. (Paper No. 82017), London: Civil Aviation Authority.

[10] Gaver, W.W. (1986). Auditory icons: using sound in computer interfaces. *Human-Computer Interaction*, 2, 167-177.

[11] Gaver, W.W. (1989). The SonicFinder: An interface that uses auditory icons. *Human-Computer Interaction*, 4(1).

[12] Gaver, W.W., Smith, R. B., & O'Shea, T.(1991). Effective sounds in complex systems: the arkola simulation. *Proceedings of the CHI '91 Conference on Human Factors in Computer Systems.* New York: ACM, pp. 85-90.

[13] Edwards, A. (1989). Soundtrack: an auditory interface for blind users. *Human-Computer Interaction*, 4(1), Spring.

[14] Kramer, G. (Ed.) (1993). *Proceedings of the First International Conference on Auditory Display*. Reading, Massachusetts: Addison-Wesley.

[15] Kramer, G. (1993). An introduction to auditory display. In Kramer, G. (Ed.) (1993). *Proceedings of the First International Conference on Auditory Display*. Reading, Massachusetts: Addison-Wesley.

[16] Kramer, G. (1993). Some organizing principles for representing data with sound. In Kramer, G. (Ed.) (1993). *Proceedings of the First International Conference on Auditory Display*. Reading, Massachusetts: Addison-Wesley.

[17] Bly, S. (1982). *Sound and Computer Information Presentation*. Unpublished Doctoral thesis (UCRL-53282), University of California, Davis, CA.

[18] Bly, S. (Ed.) (1985). Communicating with sound. *Proceedings of CHI '85 Conference on Human Factors in Computer Systems*. New York: ACM, pp. 115-119.

[19] Lunney, D. & Morrison, R.C. (1981). High technology laboratory aids for visually handicapped chemistry students. *Journal of Chemical Education*, 58 (3), 228-231.

[20] Lunney, D., Morrison, R.C., Cetera, M.M., Hartness, R.V., Mills, R.T., Salt, A.D., & Sowell, D.C. (1983). A microcomputer-based laboratory aid for visually impaired students. *IEEE Micro*, 3(4), 19-31.

[21] Lunney, D., Morrison, R.C., Sowell, D.C., and Mills, R.T. (1984). Use of complex audio signals to present multivariate data to visually handicapped students. Unpublished paper, East Carolina University, March 1984.

[22] Morrison, R.C., & Lunney, D. (1987). Using high technology to develop aids for the visually impaired scientist. Abstracts of the American Chemical Society's 193rd National Meeting. Denver, Colorado.

[23] Mezrich, J.J., Frysinger, S., and Slivjanovski, R. (1984). Dynamic representation of multivariate time series data. *Journal of the American Statistical Association* 79, 34-40.

[24] Smith, S., Bergeron, R.D., and Grinstein, G.G. (1990). Stereophonic and surface sound generation for exploratory data analysis. *Proceedings of the CHI '90 Conference on Human Factors in Computer Systems.* New York: ACM, pp. 125-132.

[25] Blattner, M.M., Sumikawa, D.A., & Greenberg, R.M. (1989). Earcons and icons: Their structure and common design principles. *Human- Computer Interaction*, 4, 11-44.

[26] Batteau, D.W. (1967). The role of the pinna in human localization. *Proceedings of the Royal Society of London*, B168, 158-180.

[27] Durlach, N.I., Rigopulos, A., Pang, X.D., Woods, W.S., Kulkarni, A, Colburn, H.S., & Wenzel, E.M. (1992). On the externalization of auditory images. *Presence: Teleoperators and Virtual Environments*, Spring, 251-257.

[28] Blauert, J. (1983). *Spatial Hearing: The Psychophysics of Human Sound Localization* (J.S. Allen, Trans.). Cambridge, MA: MIT Press.

[29] Mershon, D. H., & King, L.E. (1975). Intensity and reverberation as factors in the auditory perception of egocentric distance. *Perception and Psychophysics*, 18, 409-415.

[30] Doll, T.J., Hanna, T.E., & Russotti, J.S. (1992). Masking in three-dimensional auditory displays. *Human Factors*, 34(3), 255-265.

[31] Grantham, D.W. (1986). Detection and discrimination of simulated motion of auditory targets in the horizontal plane. *Journal of the Acoustical Society of America*, 79, 1939-1949.

[32] Strybel, T.Z. (1988). Perception of real and simulated motion in the auditory modality. *Proceedings of the Human Factors Society 32nd Annual Meeting.* Santa Monica, CA: Human Factors Society, 76-80.

[33] Strybel, T.Z., Manligas, C.L., & Perrott, D.R.(1992). Minimum audible movement angle as a function of the azimuth and elevation of the source. *Human Factors*, 34(3), 267-275.

[34] Warren, D.H., Welch, R.B., & McCarthy, T.J. (1981). The role of visual-auditory "compellingness" in the ventriloquism effect: Implications for transitivity among the spatial senses. *Perception and Psychophysics*, 30, 557-564.

[35] Foster, S. (1992). The Convolvotron: Real-time demonstration of reverberant virtual acoustic environments. *Journal of the Acoustical Society of America*, 92 (4 -2), 2376.

[36] Stuart, R. (1992). Virtual reality: Directions in research and development. *Interactive Learning International*, 8 (2), (April), 95-100.

[37] Perrott, D.R., Sadralodabai, T., Saberi, K., & Strybel, T.Z. (1991). Aurally aided visual search in the central visual field: Effects of visual load and visual enhancement of the target. *Human Factors*, 33, 389-400.

[38] Doll, T.J., & Hanna, T.E. (1989). Enhanced detection with bimodal sonar displays. *Human Factors*, 31, 539-550.

[39] Sorkin, R.D., Wightman, F.L., Kistler, D.S., & Elvers, G.C. (1989). An exploratory study on the use of movement-correlated cues in an auditory head-up display. *Human Factors*, 31, 161-166.

[40] Calhoun, G.L., Valencia, G.., & Furness, T.A. (1987). Three-Dimensional auditory cue simulation for crew station design/evaluation. *Proceedings of the Human Factors Society 31st Annual Meeting*, Santa Monica, CA: Human Factors Society, 1398- 1402.

[41] Plomp, R. (1977). Acoustical aspects of cocktail parties. *Acustica*, 38, 186-191.

[42] Urdang, E., & Stuart, R. (1992). Orientation enhancement through integrated virtual reality and geographic information systems. *Virtual Reality and Persons with Disabilities*, California State University, Northridge, March 18-21, 55-62.

[43] Wenzel, E.M.(1992). Localization in virtual acoustic displays. *Presence: Teleoperators and Virtual Environments*, (Winter), 80-107.

[44] Calhoun, G.L.,Janson, W.P., & Valencia, G. (1988). Effectiveness of three-dimensional auditory directional cues. *Proceedings of the Human Factors Society 32nd Annual Meeting*, Santa Monica, CA: Human Factors Society, pp. 68-72.

[45] Rousseau, D. (1993). 3-D displays and controls for sonar operators. To appear in *Virtual Reality Systems: Applications, Research & Development* 1 (2).

[46] Loomis, J.M., Hebert, C., & Cicinelli, J.G. (1990). Active localization of virtual sounds. *Journal of the Acoustical Society of America*, 88, 1757-1764.

[47] Wightman, F.L., & Kistler, D.J. (1989). Headphone simulation of free-field listening I: Stimulus synthesis. *Journal of the Acoustical Society of America*, 85, 858-867.

[48] Wenzel, E.M., Wightman, F.L.,& Kistler, D.J.(1991). Localization with non-individualized virtual acoustic display cues. *Proceedings of the CHI '91 Conference on Human Factors in Computer Systems*. New York: ACM, pp. 351-359.

[49] Stuart, R. (1992). Virtual auditory worlds: an overview. *VR Becomes a Business: Proceedings of Virtual Reality '92*, Westport, CT: Meckler, pp. 144-166.

[50] Durlach, N.I. (1991). Auditory localization in teleoperator and virtual environment systems: Ideas, issues, and problems. *Perception*, 20, 543-554.

[51] Lenat, D.B., and Guha, R.V. (1991). Ideas for applying Cyc. MCC Technical Report Number ACT-CYC-407-91, Austin TX, November.

16

Contibutors

Mark Bajuk

1631 Denniston Avenue, Pittsburgh, PA 15217; mbak@andrew.cmu.edu.

Mark Bajuk received a BS in Electrical Engineering and BFA in Painting from the University of Illinois at Urbana-Champaign, and has worked as a Scientific Animator with the Visualization Group at NCSA. As a member of this group, he worked on animations that have received awards and recognition from conferences such as SIGGRAPH, NICOGRAPH (Japan), NCGA, Imagina (Monte Carlo), American Film and Video Festival, and the Computer Graphics Film Festival (London). These animations have also been exhibited in museums including the Beaubourg National Museum of Art (Paris), Cite des Sciences et de l'Industrie (Paris), Caixa de Pensions (Barcelona), Power Plant (Toronto), Modern Art Museum (San Francisco), and the Hong Kong Museum of Science and Technology. He is currently pursing an MFA in computer art at Carnegie Mellon University and teaching 3-D computer graphics to artists and designers.

Nalini Bhushan

Department of Philosophy, Smith College, Northampton, MA 01063; nbushan@smith.smith.edu.

Nalini Bhushan is Assistant Professor of Philosophy at Smith College. She received her doctorate from the University of Michigan, Ann Arbor in 1989 and has been teaching at Smith since then. Her areas of interest are philosophy of language and philosophy of mind. She is currently doing work on metaphor and collaborative research on texture as a visual property of objects.

Michael Heim

Education Foundation of the Data Processing Management Asscoiation, 2041 San Anseline, Long Beach, CA 90815;mheim@csulb.edu.

Michael Heim, Ph.D., is the Virtual Reality consultant for the Education Foundation of the Data Processing Management Association. He organized and chaired four national conferences on VR for top decision-makers in Washington, D.C. He is the author of the book *The Metaphysics of Virtual Reality* (Oxford University Press, 1993) and of the book *Electric Language*: *A Philosophical Study of Word Processing* (Yale University Press, 1989). He has lectured on VR at universities in Europe and the United States, and his ideas about cyberspace have appeared in newspapers as well as in academic books such as *Cyberspace: First Steps* (MIT Press, 1991). He was a Fulbright scholar for three years in Europe, and is now a freelance philosophy professor living in Long Beach, California.

Thomas Hubbard

School of Journalism, The Ohio State University, 242 W. 18th Avenue, Columbus, OH 43210; (614) 292-9633;hubbard.1@osu.edu.

Tom Hubbard was a local television director for ten years in Norfolk, Va. and Atlanta, Ga. He was a newspaper photo journalist for twelve years at the Cincinnati Enquirer, where he won twenty two Ohio awards for photo journalism and two national awards, including one for best photo story coverage of the Senate Watergate hearings. He has taught photo journalism at The Ohio State University School of Journalism since 1983. His occasional freelance assignments have appeared in many national magazines including Time, People Weekly, US Magazine, The New York Times, Business Week, Fortune, and Family Circle. Hubbard has edited the Viscom Newsletter for an Association for Education in Journalism/Mass Communications (AEJMC) Division and the ONPA Newsletter for the Ohio News Photographers Association.

Beverly J. Jones

School of Architecture and Allied Arts, University of Oregon, Eugene, Oregon 97403; beverlyjones @aaa.uoregon.edu.

Dr. Beverly J. Jones is a department head and director of graduate research in the School of Architecture and Allied Arts at the University of Oregon. She also is curriculum consultant and research supervisor for the Applied Information Management Program (AIM) of the University of Oregon. She has worked in applications of computers and information science to research and practice in the arts and education since the mid-1970s.

Alyce Kaprow

The•New•Studio, Newton, MA 02166; Voice: (617) 969-0288; Fax: (617) 965-6207.

Alyce Kaprow, president of THE•NEW•STUDIO, is a consultant to developers and users of computer graphics, and in graphics interface design; and an illustrator and graphic designer. She has worked as a designer for private industry, government and education; as a professional photographer; and has taught photography, design media technology and computer graphics for the past eighteen years. She has chaired the Designing Technology show for the 1993 SIGGRAPH Conference, was the Technical Panels Chair for 1990 SIGGRAPH Conference, and is currently on the Executive Committee of SIGGRAPH as Director of Communications and the advisory board for electronic publications for the ACM. She is contributing editor to various publications, including Pre-Magazine, Magazine Design & Production.

Ms. Kaprow received a BFA in Design from Syracuse University, an MFA in Photography from California Institute of Arts, and Certificate Degrees from Agfa-Gevaret Technikum and Linhof Institute, both in Munich, Germany. She has done post-graduate research in computer graphics and media technologies at the Visible Language Workshop, MIT.

Xia Lin

School of Library and Information Science, University of Kentucky, Lexington, KY 40506; Lxia00 @aix3090b.uky.edu

Dr. Xia Lin is Assistant Professor in the School of Library and Information Science. He has a Ph.D. from University of Maryland. His research interests include information retrieval, human-computer interface, hypertext, and neural networks. He has published several articles in ACM, IEEE, and ASIS publications.

John Loustau

Department of Mathematics and Statistics, Hunter College of the City University of New York, 695 Park Avenue, NY, NY 10021.

John Loustau is an Associate professor, Department of Mathematics and Statistics, Hunter College of the City University of New York where he directs a masters program in geometric modeling. Additionally, he is an independent consultant and lecturer to the aerospace and health industries in the United States and Asia, specializing in computer graphics. His published works include articles in mathematics, computer graphics and a text book, *Linear Geometry with Computer Graphics* published by Marcel DekkerInc. (1993). Dr. Loustau is a member of ACM/SIGGRAPH, NYC/SIGGRAPH, IEEE, AMS and MAA. He received the BA degree in mathematics from Oregon State University, Corvallis, and the PhD degree in mathematics from the University of California, Santa Barbara.

Francis T. Marchese

Computer Science Department, Pace University, 1 Pace Plaza, New York, NY 10038-1502; Voice:(212) 346-1803; Fax: (212) 346-1863; marchesf @PACEVM. bitnet.

Francis T. Marchese received his Ph.D. in Chemistry from the University of Cincinnati in 1979. He was an National Institutes of Health Postdoctoral Research Fellow from 1978-1983 at Hunter College of CUNY. He is Professor of Computer Science at Pace University where his research encompasses computer graphics, scientific visualization, and computer simulation of self-organizing systems. In 1992, he received the Kenan Award for Teaching Excellence and is presently writing texts on molecular visualization and graphics that will be published by Telos/Springer-Verlag.

Marc De Mey

University of Ghent, Biandijnberg 2, 9000 Gent, Belgium; Voice: 091 64 39 52; Fax: 32 91 64 41 97.

Marc De Mey is professor in psychology and history of science at the University of Ghent in Belgium. With a background in cognitive psychology obtained during a stay at Harvard's Center for Cognitive Studies, he applies ideas and methods based upon Piagetand AI in the analysis of scientific discovery. His case study of Harvey's discovery of circulation is report in his book The Cognitive Paradigm (originally Reidel 1982, 1984 paperback) recently republished in a paperback edition by The University of Chicago Press with a new introduction (1992). He is currently involved in a cognitive science project on linear perspective, more in particular the links between late medieval optics and the discovery of perspective in art. Marc De Mey has been a member of the council of EASST (European Society for Social Study or Science) and SSSS (Society for Social Study or Science).

A. Ravishankar Rao

IBM T.J.Watson Research Center, Yorktown Heights, NY 10598-0704; (914) 945 3553; rao@watson.ibm.com.

A. Ravishankar Rao received his doctorate in computer Engineering from the University of Michigan, Ann Arbor, in 1989. He joined the IBM Thomas J. Watson Research Center, Yorktown Heights, New York, in 1990, where he is currently a research staff member. His research interests include computer vision, artificial intelligence, and automation in manufacturing. He received the Best Student Paper award at the VISION-88 Conference, held in Detroit in 1988. He is the author of the book *A Taxonomy for Texture Description and Identification*, published by Springer Verlag (1990).

Judson Rosebush

Judson Rosebush Company, 154 W. 57th Street, #826, New York, NY 10019; (212) 398-6600.

Judson Rosebush is a producer and director of computer animation, an author, and media theorist. He graduated from the College of Wooster in 1969 and received a PhD from Syracuse University in communications research. He has worked in radio and television broadcasting, sound and video production, print, and hypermedia. He completed his first computer animations in 1970 and founded Digital Effects Inc. in New York (1978-1985). He has exhibited computer generated drawings and films in numerous museum shows and his computer drawings have been reproduced in hundreds of magazines and books.

Rosebush is the co-author of *Computer Graphics for Designers and Artists*, published by Van Nostrand Reinhold Co., and is currently completing a book on Computer Animation. He is the author of the serialized Pixel Handbook. During the past two years he has co-authored and directed television programs on Volume Visualization, HDTV and the Quest for Virtual Reality, and Visualization Software, written "The Proceduralist Manifesto" a tutorial on Using APL for Computer Graphics Notation, and programmed a HyperCard controlled videodisc system. Projects in production include Return to Trinity, a multimedia documentary on the world's first nuclear bomb test, and multimedia titles for publication on CD/ROM.

Michael J. Shiffer

Department of Urban Studies and Planning, Massachusetts Institute of Technology, 77 Massachusetts Avenue, Room 9-514, Cambridge, MA 02139; Voice: (617) 472-0200; Fax: (617) 253-3625; mshiffer@athena.mit.edu.

Michael J. Shiffer is a Postdoctoral Associate and Lecturer at the Massachusetts Institute of Technology's Department of Urban Studies and Planning. He is an active professional speaker and author on the use and development of hypermedia technologies in the area of urban planning and decision-making. Shiffer's current research interest is multimedia and electronic communication to facilitate group planning processes. He has consulted with several public and private agencies including the U.S. Army, the U.S. Department of Transportation, the World Bank, and the City of St. Louis. Shiffer received his Ph.D. in Regional Planning and Master's of Urban Planning from the University of Illinois at Urbana-Champaign.

Rory Stuart

Intelligent Interfaces Group/Artificial Intelligence Laboratory, NYNEX Science and Technology Inc., White Plains, NY 10604; (914) 644-2362; stuart @nynexst.com.

Rory Stuart is a member of the technical staff in the Computer/ Human Interaction Laboratory at NYNEX Science and Technology, Inc., where he has designed and evaluated human-computer interfaces since 1987. Currently, he heads the Virtual Reality Project at NYNEX with research interests in virtual acoustic displays and sonification. In addition, he has designed phone-based interfaces and performed human factors studies of multimedia communications systems. He is editor-in-chief of SIG-Advanced Applications Virtual Reality Systems: Applications, Research.

Barbara Tversky

Department of Psychology, Stanford University, Stanford, CA 94305; bt@waldron.stanford.edu.

Barbara Tversky received a PhD in Cognitive from the Psychology University of Michigan in 1969. She has served at Hebrew University from 1967-1977 and has been at Stanford University since 1978. Her interests include visual/spatial thinking and memory categorization.

Jong-ding Wang

Department of Mathematics and Statistics, Hunter College of the City University of New York, 695 Park Avenue, New York, NY 10021.

Jong-ding Wang is a graduate student, Department of Mathematics and Statistics, Hunter College of the City University of New York. Mr. Wang has received the AA degree in electrical engineering from Kwang-Wu Junior College of Technology, Taipei, Taiwan, ROC, the BA degree in computer science from Hunter College CUNY. Mr. Wang is a member of ACM/SIGGRAPH and MAA.

Robert S. Williams

Department of Computer Science, Pace University, 1 Pace Plaza, New York, NY 10038-1502; (212) 346-1336; williams@PACEVM.bitnet

Robert S. Williams received the PhD in Computer Science from the University of Massachusetts in 1990. He spent the following year working with the University's Natural Language Group, and is currently an Assistant Professor at Pace University. He has published in the areas of machine learning, case-based reasoning, and natural language processing, and is also interested in programming languages for artificial intelligence. In addition, Dr. Williams has played guitar since 1970, and has recently been heard performing in the New York Guitar Project.